Ethics and Biomedical Engineering

This book is about the interaction between biomedical engineering and ethics during emergencies, such as low-resource settings and the COVID-19 pandemic. It addresses the issues between the universalism of human rights, ethical principles, and regulatory standards of biomedical devices in the context of emergencies.

Ethics and Biomedical Engineering: Facing Global Health Emergencies connects biomedical engineering and ethics with particular regard to emergency context such as low-income countries and the COVID-19 pandemic. It examines how the COVID-19 crisis exposed gaps in access to healthcare, ignited debates about resource allocation, and highlighted the importance of patient privacy. The book presents case studies conducted in Africa and the role of the biomedical engineer (and more generally the scientist) during a pandemic or other health emergency. The book also addresses the way in which the pandemic has been addressed in low-income contexts. Finally, it also explains the need for an interdisciplinary approach between scientists, ethicists, and policymakers to improve outcomes in the future.

The book is intended for undergraduate and graduate students in bioengineering. It would also be useful for policy makers and medical professionals that could be faced with ethical dilemmas in times of crisis.

Ethics and Biomedical Engineering

Facing Global Health Emergencies

Leandro Pecchia, Alessia Maccaro, and
Kallirroi Stavrianou

CRC Press
Taylor & Francis Group
Boca Raton London New York

CRC Press is an imprint of the
Taylor & Francis Group, an **informa** business

Designed cover image: © Shutterstock

First edition published 2025
by CRC Press
2385 NW Executive Center Drive, Suite 320, Boca Raton FL 33431

and by CRC Press
4 Park Square, Milton Park, Abingdon, Oxon, OX14 4RN

CRC Press is an imprint of Taylor & Francis Group, LLC

ISBN: 978-1-032-56610-8 (hbk)
ISBN: 978-1-032-56609-2 (pbk)
ISBN: 978-1-003-43634-8 (ebk)

DOI: 10.1201/9781003436348

Typeset in Times
by Newgen Publishing UK

Contents

About the Authors

Leandro Pecchia is Chair of Biomedical Engineering at the University Campus Bio-Medico in Rome, Italy and the Rector Delegate for Internationalization and Sustainable Development Goals. In 2021, the Italian Minister of Health appointed him as one of the three medical device experts of the Technical and Clinical Committee of the Ministry of Health. From 2020 to September 2023, he worked as Innovation Manager for the World Health Organization (WHO), Emergency Program (WHE) Infection Prevention and Control (IPC) Unit. Prof Pecchia has authored more than 200 peer-reviewed papers on journals, books, and conferences in the fields of Artificial Intelligence, medical devices, biomedical signal processing, Health Technology Assessment (HTA), Internet of Things, and machine learning applied to active and healthy ageing and management of chronic diseases. Prof Pecchia is the Secretary General of the International Federation of Medical and Biological Engineering (IFMBE), an NGO in official relations with the WHO, and the Past President of the European Scientific Society of Biomedical Engineering.

Alessia Maccaro is a Research Fellow in Moral Philosophy and Bioethics for the School of Engineering of the University of Warwick, United Kingdom. She is an expert in interdisciplinary research on the ethical and regulatory aspects of healthcare technologies, in particular in low-resource settings and in emergency situations such as pandemics. Currently she co-leads a work package on the pilot management for the Horizon 2020 ODIN Project on Smart Hospitals. She obtained the National Academic Qualification as Associate Professor of Moral Philosophy for the SS M-FIL/03 in 2020. She is a member of several professional bodies including the Italian Society of Moral Philosophy (SIFM), Interuniversity Center of Bioethical Research (CIRB – University of Naples Federico II), and serves on the Scientific Advisory Board of the Laboratory "Ethics, Bioethics and citizenship" (EBC), University of Naples Federico II.

Kallirroi Stavrianou is an experienced Postdoctoral Researcher with a demonstrated history of working in the higher education industry and skilled in Medical Devices, Health Technology Assessment (HTA), and Health Technology Management. She is a strong research professional with a Doctor of Philosophy (PhD) and Master of Science (MSc) focused on Clinical Engineering, and a Bachelor in Physics from the University of Patras. She is the manager of the Global Clinical Engineering Alliance and Global Clinical Engineering Journal, Board member of the Clinical Engineering Division of IFMBE, former member of the Health Technology Assessment Division of IFMBE, and chairs the IFMBE Publicity Committee. Dr. Stavrianou is also a member of the Hellenic Society for Biomedical Technology (ELEVIT).

Introduction

The world has never been more interconnected, and this interconnectedness has brought both incredible opportunities and daunting challenges. One of the most pressing challenges we face today is global health emergencies, such as the COVID-19 pandemic, which test the limits of our healthcare systems, our scientific knowledge, and our ethical frameworks.

Biomedical engineering plays a crucial role in addressing these challenges. From developing new diagnostic tools to designing innovative medical devices and healthcare infrastructure, biomedical engineers are at the forefront of the fight against global health emergencies. However, the rapid pace of technological advancement and the complexities of these emergencies raise a multitude of ethical questions.

The COVID-19 pandemic serves as a stark reminder of the vulnerabilities exposed during global health emergencies. The rapid spread of the virus exposed gaps in healthcare infrastructure, highlighted inequalities in access to medical technologies, and ignited complex ethical debates surrounding resource allocation, patient privacy, and vaccine distribution. This book dissects these challenges through a series of in-depth chapters, offering crucial insights from biomedical engineers and ethicists who were on the frontlines of the pandemic response.

Ethics and Biomedical Engineering: Facing Global Health Emergencies delves into these critical questions. It transcends the specific context of COVID-19, offering a vital toolkit for navigating future global health emergencies. By dissecting the ethical dilemmas of the present, it equips us with the knowledge and foresight to create a more just and equitable healthcare landscape for all. As we emerge from the shadow of pandemics, let us not forget the lessons learned – lessons etched in the lines of these articles, and a testament to the enduring power of ethics and biomedical engineering to face the challenges of a vulnerable world.

The book brings together a collection of thought-provoking articles that explore the ethical implications of biomedical engineering in the context of global health emergencies and is divided into three sections:

DOI: 10.1201/9781003436348-1

Part 1: Biomedical Engineering and Ethics Face COVID-19 Pandemic

Part 2: Cases Studies in Low-Resource Settings

Part 3: Ethics and Regulation

Part 1 examines the ethical considerations surrounding the development and deployment of medical technologies during the COVID-19 pandemic. The chapters in this section address issues such as the equitable distribution of medical resources, the ethical implications of rapid prototyping and innovation, the privacy concerns associated with digital health technologies, and the role of ethics in scientific research during a crisis.

Part 2 shifts the focus to low-resource settings, where the challenges of global health emergencies are often exacerbated by limited resources and infrastructure. The articles in this section explore issues such as the need for context-specific ethical frameworks, the importance of community engagement and social dynamics, the development of medical devices and healthcare systems that are resilient to resource constraints, and the ethical implications of global health interventions in low-resource settings.

Part 3 examines the regulatory frameworks that govern biomedical engineering and healthcare. The articles in this section address issues such as the adequacy of existing regulations in the face of global health emergencies, the need for international collaboration and harmonization of regulations, the ethical challenges of balancing innovation with safety and efficacy, and the role of ethics in ensuring the responsible development and deployment of medical technologies.

The overarching aim of this book is to provide a comprehensive and critical analysis of the ethical considerations that must guide biomedical engineering during global health emergencies. By exploring diverse case studies, offering practical frameworks, and engaging with ethical debates, this book seeks to equip readers with the knowledge and tools to navigate the complexities of this critical field.

This book is not just for academics and experts. It is for anyone who is concerned about the future of healthcare and the ethical implications of our technological advancements. It is for anyone who believes that we have a responsibility to ensure that the benefits of biomedical engineering are shared equitably and ethically around the world. We hope you will join us on this journey as we explore the complex intersection of ethics, biomedical engineering, and global health emergencies.

We believe that this book will be a valuable resource for anyone who is interested in the ethical implications of biomedical engineering in the face of global health emergencies, and we hope that it will spark important conversations about the future of healthcare and the ethical challenges we face. Moreover, we believe that by working collaboratively and ethically, biomedical engineering can play a transformative role in building a more resilient and equitable world, prepared to face the ongoing and emerging health challenges of the 21st century.

Statement

This book, apart from the unpublished chapters, has the inclusion of eight articles on Ethics and Biomedical Engineering, published in open access high impact journals during the first months of the pandemic. All eight articles are licensed under a Creative Commons Attribution 4.0 International License which permits use, sharing, adaptation, distribution, and reproduction in any medium or format (https://creative commons.org/licenses/by/4.0/). There were no modifications made in the following articles, which are presented in three different sections:

I – Biomedical Engineering and Ethics Face Covid-19 Pandemic

- A. Maccaro, D. Piaggio, C. Dodaro, L. Pecchia, *Biomedical engineering and ethics: reflections on medical devices and PPE at the time of Covid-19*, in «BMC Medical Ethics», (2021) 22:130 https://doi.org/10.1186/s12910-021-00697-1, https://rdcu.be/cyAIH
- A. Maccaro, D. Piaggio, S. Pagliara, L. Pecchia, *The role of ethics in science: a systematic literature review from the first wave of COVID-19*, in «Health and Technology» (Springer), 03/06/2021: https://link.springer.com/content/pdf/10.1007/s12553-021-00570-6.pdf

II – Cases Studies in Low-Resource Settings

- A. Maccaro, D. Piaggio, M. Vignigbé, A. Singl, L. Pecchia, *COVID-19 preparedness and social dynamics in a Sub-Saharan Africa country, Benin*, in «Health Promotion International», Volume 37, Issue 4, August 2022, daac105, https://doi.org/10.1093/heapro/daac105 (https://academic.oup.com/heapro/article/37/4/daac105/6674369?searchresult=1)
- A. Maccaro, D. Piaggio, I. B. Oronti, M. Vignigbé, A. Gbokli, R. Hougnihin, L. Pecchia, *Social engagement in the fight against COVID-19 in the urban and peri-urban areas of Cotonou (Benin, Sub-Saharan Africa): acceptability of the vaccination and tracking program*, in «Frontiers in Medicine», 3 June 2022: www.frontiersin.org/articles/10.3389/fmed.2022.857890/full (https://doi.org/10.3389/fmed.2022.857890).
- D. Piaggio, R. Castaldo, M. Cinelli, S. Cinelli, A. Maccaro, L. Pecchia, *A MCDA framework for designing medical devices resilient to low-resource settings: an application for low- and middle-income countries (LMICs)*, in «Globalization and Health» 17, 64 (2021) https://globalizationandhealth.biomedcentral.com/articles/10.1186/s12992-021-00718-z
- L. Di Pietro, D. Piaggio, I. Oronti, A. Maccaro, R.C. Houosseuvo, D. Medenou, C. De Maria, L. Pecchia, A. Ahluwalia, A Framework for Assessing Healthcare Facilities in Low Resource Settings: Field Studies in Benin and Uganda, in «Journal of Medical and Biological Engineering» (Springer), (24 June 2020) 40:526-534 (DOI: https://doi.org/10.1007/s40846-020-00546-3)

III – Ethics and Regulation

- L. Pecchia, D. Piaggio, A. Maccaro, C. Formisano, E. Iadanza, The inadequacy of regulatory frameworks in time of crisis and in Low-Resource Settings: Personal Protective Equipment and COVID19, in «Health and Technology» (Springer), 2 May 2020: https://link.springer.com/article/10.1007/s12553-020-00429-2 (DOI: https://doi.org/10.1007/s12553-020-00429-2)
- A. Maccaro, D. Piaggio, S. Leesurakarn, S. Sekalala, S. M. Rai, L. Pecchia, On the universality of Medical Device Regulations: the case of Benin, in «BMC Health Services Research», (2022) 22:1031, https://doi.org/10.1186/s12913-022-08396-2

Chapter 1

Biomedical Engineering and Ethics Face COVID-19 Pandemic

Biomedical Engineering and Ethics: Reflections on Medical Devices and Personal Protective Equipment During the First Wave of COVID-19[1]

Background

Since early 2020, the current SARS-CoV-2 pandemic has challenged all the fields of knowledge, increasing the need for their interconnection. Medicine, science, politics, and more specialized sectors such as biomedical engineering (BME), faced crucial ethical issues, which can no longer be underestimated.

Biomedical engineers design medical devices (MDs), raising many ethical dilemmas in ordinary times, which become compelling during such a crisis. The authors of this manuscript had the privilege of different points of view thanks to the President of the European Alliance of Medical and Biological Engineering and Science (EAMBES) and Secretary General of the International Union for Physical and Engineering Science in Medicine (IUPESM).

Three pivotal themes emerged in the global BME community:

1. The dilemma of identifying criteria for the allocation of MDs
2. Responsibilities of science and technology
3. Inadequacy of regulations and norms, which lack universality

This manuscript does not follow the traditional structure of scientific papers (e.g., methods, results etc.), rather it is a critical analysis, revolving around the three above-mentioned pillars. The first pillar focuses on the surfacing of ethical dilemmas in times of pandemic (e.g., scarce resource allocation), delving into examples from Italy, where the new decision-making criteria often clashed against the existing constitutional and moral principles during the first wave.

The second pillar retrospectively reports how the MD sector was affected by the current pandemic, touching on hazardous amateur attempts of the general public to face the scarcity of resources and the urgent needs, and on the crisis-related challenges

DOI: 10.1201/9781003436348-3

that surfaced for manufacturers, exacerbated by the lack of dialogue with decision-makers. In the same pillar, the theoretical debate among science and politics is addressed, referring the case of the "intended use" of a MD, from two different philosophical perspectives, i.e., substantialism and utilitarianism. The former underlines the fundamentality of substances as ontological categories (1), suggesting that the intended use of a MD should always be respected. The latter stresses the importance of maximizing the overall good, authorizing different uses, depending on the relativistic utility or, in exceptional times of crisis, on the emergency.

This invites the reader to a subsequent reflection on the inadequacy of the existing regulations on MDs, in the third pillar. In respect to this, the manuscript proposes to offer a hermeneutic perspective close to the situational ethics that authorizes negotiations and mediations between the generality of principles and norms, and the specificity of the context.

Overall, the ethical considerations made in this manuscript should be considered a valuable lesson for the future of crisis management. If and only if ethics and bioethics will be considered as effective support for science and scientists (doctors, biomedical engineers, etc.), the Cartesian separation of knowledge (2) could be overcome, establishing an interdisciplinary dialogue that involves peoples and emphasizes the public relevance of such issues.

In this way, such dilemmas could be anticipated by establishing a framework that could provide guidance and appropriate methodology to address arising urgent issues, without having to resort to specialists for questions that concern everyone and need a multidisciplinary approach.

Allocation of Medical Devices: Clinical and Ethical Principles

Due to limited resources, decision-makers have had to compromise among all the potential useful interventions. Also in the past ten years, the most public national health systems were massively privatized, resulting in a significant reduction of prevention services and a great reduction of intensive care unit (ICU) beds. The rapid spread of the SARS-CoV-2 resulted in an unprecedented need of sub-intensive and ICU beds, which overcame the capacity of the most advanced national healthcare services. Within a few weeks, the available resources (i.e., MDs, doctors, nurses) proved to be insufficient to cover the care needs of the multitude of COVID-19 patients, beyond the ordinary needs of other patients. Consequently, doctors and healthcare structures ended up pondering and making difficult ethical choices in a short time and identifying priority principles that could guide them.

National (3–6) and international (7–9) ethics committees, scientific societies (10, 11), and experts (12–15) soon expressed their opinion on the matter (16), identifying requirements that respect human dignity and fundamental ethical principles, enshrined in the Charters of both National and International Rights (17).

The case of Italy, the first country significantly affected outside China, is emblematic. According to the Italian Constitution, health is a "fundamental right of the individual" and a "collective interest" (article 32). In addition, article 2 recognizes the personalist principle and the duty of solidarity, and article 3 establishes the principle of equality. Accordingly, the law founding the Italian National Health Services (NHS) (n. 883/1978) prescribes that care must be ensured according to the principles of universality, equality, and fairness.

Leveraging on these fundamental principles, the Italian National Bioethics Committee (18) remarked the need to allocate MDs and other resources based only on a clinical criterion, and without considering criteria such as age, gender, social attributes, ethnicity, disability, and costs, in compliance with the principles of "justice, equity, solidarity". The envisaged method was that of "triage in a pandemic emergency", based on what the World Health Organization (WHO) defines as "preparedness" (19) as a premise, and on two key concepts, i.e. "clinical appropriateness" and "actuality", identified by the healthcare professional on the basis of clinical criteria.

This point of view, which values the person and opposes attempts of objectification in a series of "pre-established criteria" (except for the point of view of the clinician), is well expressed by J. Habermas. In an interview with the newspaper Le Monde, Habermas underlined the inadequacy, in moral terms, of an "objective quantification" of patients and insisted on the essential issue of the "recognition" of the individual: "when addressing a second person (you, you), the other person's self-determination must either be respected or denied, that is, either accepted or ignored" (The Only Cure Is Solidarity from (20)).

On the other hand, the Italian society of anesthesiology (Italian Society for Anesthesia Analgesia Resuscitation and Intensive Care [SIAARTI]) introduced the identification of an age limit for accessing intensive care in case of necessity (21). SIAARTI's document explained that COVID-19 created a scenario in which criteria for accessing the ICU may be needed beyond the clinical appropriateness and proportionality of care, but also in distributive justice and the appropriate allocation of limited healthcare resources, which means privileging those with the "greatest life expectancy" (22). Even if the choice of whom to admit to treatment is a terrible reality, it must be discussed by ethicists and bioethicists to further identify solutions that support medical doctors in taking these decisions. "This is not to deny their clinical authority and responsibility, but rather to urge a commitment to give such question public relevance" so that in the future the "public health perspective" (23) to follow is clear from the beginning.

Responsible Thinking, Responsible Actions, Responsible Silence

COVID-19 created a global lack of essential MDs (e.g., pulmonary ventilators) and personal protective equipment (PPE, such as masks and respirators) (24). This led to an unprecedented amount of do-it-yourself (DIY) solutions, which were fomented on media worldwide. Consequently, ordinary individuals started producing PPE at home with 3D printers and everyday materials, and manufacturers converted their production facilities to develop MDs and PPE.

Unfortunately, although very admirable, this approach is not feasible in critical sectors such as MDs or PPE, which require postgraduate education, years of experience, and deep knowledge of relevant international standards and norms in order to ensure appropriate levels of safety, efficacy, and resilience. Thus, only seven manufacturers in the world are producing pulmonary ventilators. In fact, during a pandemic, we do not only need to ensure the usual standard of quality, but we should also consider making those devices more resilient, because, if hospitals fail, we will need to safely operationalize these devices in field hospitals, tent-like structures, and any other relevant setting. Therefore, experience, safe-by-design approaches, and the knowledge of additional standards (e.g. military standards) become relevant too.

Minimal scientific evidence exists on how harmful this DIY wave has been, but a few facts can be clearly reported referencing major newspapers. In mid-March 2020, manufacturers were called upon to help to tackle the MDs and PPE crisis.[2,3] Many responded, certainly moved by the noblest principle and willingness to help. Unfortunately, learning to manufacture complicated and highly regulated pulmonary ventilators cannot be done in a few weeks. By mid-April 2020, new productions of ventilators were stopped in Spain,[4] UK suspended orders of BlueSky ventilators,[5] and France followed.[6] Once again, a virtuous example came from Italy, where the only Italian manufacturer of ventilators (i.e. Siare Engineering International Group Ltd.) was supported by the Italian Government, which offered 25 highly specialized army engineers, by the former FIAT (now FCA), who supported producing electromechanic components, in addition to Ferrari providing electronic components. With this collaborative effort, Siare increased the production of top-of-the-range ventilators from 160 to 500 units per month, respecting the highest quality standards. In conclusion, the rise of useless and potentially harmful DIY approaches to PPE and MDs could have been easily avoided at the start of the pandemic by decision-makers initially consulting with domain experts, such as biomedical and clinical engineers.

In this regard, the belief that disciplinary competence is to be sectorized and not interconnected, very often, leads to a further separation of knowledge that is detrimental to people. In fact, it could be good practice that politicians have a solid scientific background in order to legislate about scientific matters, above all if they involve public health. On the other hand, scientists should "grovel in the dirt of the city of Romulus" (25) keeping their related studies as tangible and accessible as possible, and acquire a more solid political culture and a growing awareness of their social role. After all, the relationship between science, policy-making, and politics has been controversial since the dawns of civilization: people like Aristarchus of Samos (i.e., one of the fathers of an early Heliocentrism), censored by sectaries such as Cleanthes (i.e., the prince of stoics at that age), or like Socrates, accused, "censored", and sentenced to death for being "unorthodox" by one of the most open and democratic societies of the times, the *Polis* of Athens (26), or like Galileo Galilei, who was condemned for radically opposing to the Sacred Scriptures- and Church-approved Geocentric model, are just a few stark examples. Although much progress has been made since then, there is still an ongoing debate among a Weberian distinction between science and politics and a Habermasian and Marcusean dichotomy between the technocratic and decisionist models of scientific advice to politics. Nonetheless, it is evident that we are transitioning towards ever more present democratization of science, and not without associated risks. In fact, how can this be achieved without compromising the *epistemic quality* of knowledge (27)?

In particular, the COVID-19-related debate, involving politicians, scientists (especially biomedical engineers) and ethics experts, is based on two distinct currents of thought, referring to two different philosophical matrices, i.e. *utilitarianism* and *substantialism*. In fact, some people believe that any kind of emergency-ready response that can make up for the shortage of PPE and MDs is "better than nothing", even at the expense of the safety and efficiency normally guaranteed by the standards. This way of thinking is in line with utilitarianism's conception of maximizing happiness and overall gains for all the affected individuals. However, the "better than nothing approach" is dubious and is a well-known logical fallacy, that of the relative privation (28). This kind of fallacious way of reasoning also justifies the misuse of something that does meet the standards but was intended for completely different purposes. In fact, the intended purpose is key, at least in the world of MDs, and it is what safeguards the manufacturers in case their products fail if they are used "off label". In this case, the liability falls with the individual who misused the product in the first place (29). This concept is well portrayed by what happened in Harrow,[7,8] where some nurses, after denouncing their precarious working conditions and the lack of PPE in the fight against COVID-19, had started using bin bags as a replacement of the unavailable PPE. In this case, the beneficial objective was given priority and the collective benefit was maximized. However, the risk to the wearer had also rapidly increased to the extent that it was not possible to predict its consequences (even negative).

Conversely, other people believe that the intended purpose of an object should be respected, in line with substantialism's theories, which attribute absolute value to an idea. According to this perspective, for example, a bin bag would be designed, tested, and marketed to contain rubbish, not to protect healthcare workers from diseases (in which case the design principles and the tests will be different and stricter). Thus, these people tend towards a minimization of the risk, but, at the same time, their precautionary approach hinders the possible benefit underlying the other "less safe" alternatives. In this regard, it is necessary to recall the philosophy of Hans Jonas who, faced with an indeterminate and potential risk, the consequences of which cannot be estimated, introduced the *imperative of responsibility* in defense of future generations which is based on the precautionary principle. His "heuristics of fear" implies foresight and ability to predict and adequately assess the consequences of collective activities in contemporary societies. Such principle implies to "act so that the effects of your action are compatible with the permanence of genuine life" and, in our present choices, to "include the future wholeness of Man among the objects of your (our) will" (30).

Consequently, this responsibility goes beyond the personal one of engineers, as it also includes the responsibility they partially assume if and when they do not limit reckless or inadequately considered actions, guided by the above-mentioned utilitarian approach. In fact, the compliance with international standards and the consequent CE (European Conformity) marking does not only guarantee the quality, safety, efficiency, and efficacy of a product, but also the protection of manufacturers and users. In fact, as afore-mentioned from a general point of view, those to be blamed for the possible failure of bin bags used as PPE for the prevention of COVID-19 and the consequent infection and potential death of the healthcare workers using them are not the manufacturers of such items, but whomever decided to use this amateur substitution to other certified means of protection, and, to a certain extent, the biomedical engineers who did not respect their duty to identify the limits regarding the unintended uses.

Overall, two theoretical orientations are at the basis of these dichotomous approaches. However, in order to better frame them, it is necessary to analyze them in view of the extraordinary condition of necessity begotten by the pandemic. The dilemma revolves around the "intended use", or rather the purpose for which something (e.g., PPE or MDs) was originally designed for: on the one hand there are those who assert that, in conditions of necessity, the contingent purpose, i.e., the social functionality that overcomes the intended use, ought to be preferred despite being "off label". In fact, they firmly believe that it is preferable to maximize the current benefit while assuming an undefined risk. Although it is not easy to relate this trend to a specific current of thought, it certainly shares some points in common with utilitarianism, starting from the Benthamian one (31), if not with pragmatism (e.g., 32).

On the other hand, the position of those who consider the "intended use" or rather the intrinsic purpose for which the product was manufactured, tested, and marketed, a priority, would seem evocative of Aristotelian substantialism or eschatology. In this case, the risks are limited by compliance with the law and the relevant standards, which also guarantee the achievement of the benefits. The refusal of this immediately relieves the manufacturer and the regulator from any responsibility related to the misuse of the object, leaving every possible and unforeseeable risk to the individual.

Here lies the crux of the problem. Using what is available and certified, albeit designed for a different "intended use", seems more "reasonable" than not protecting oneself to everyone. However, we cannot refrain from asking ourselves the following questions: what is the limit within which it is possible to say, "better than nothing"? To what extent can science and policymakers put people's lives at risk in order to have a prompt, but probably unsafe answer in the wake of the "better than nothing" principle?

Regulatory Frameworks and Standards Should be Reviewed in this Regard

Beyond the DIY solutions, low-quality outputs have been affecting scientific production too. In fact, the high demand for information caused an acceleration in reporting scientific results, with many journals being overwhelmed with unprecedented numbers of papers, which challenged the capability of editors and reviewers to scrutinize articles.

The unprecedented high number of retracted papers can be a proxy for the high number of low-quality research on COVID-19. For this reason, a rapid search for papers regarding COVID-19 or SARS-CoV-2, and the previous epidemic/pandemics (i.e., SARS, MERS, and Swine Flu) as a comparison, was performed both on OvidSP and the Retraction Watch Database. Regarding COVID-19 publications, there were 124 retracted papers out of 264530, i.e., 4.68 per 10,000 papers (compared to 1.16 per 10,000 papers concerning the previous pandemics/epidemics). Although this proxy is to be taken with a grain of salt, it should be a wake-up call for further investigations. Similar levels of confusion could also be observed among scientists and experts invited by media to interpret available scientific evidence and technical guidance.

The above-mentioned issues contributed to beget and feed an *infodemic*, defined by the United Nation as "an over-abundance of information – some accurate and some not – that makes it hard for people to find trustworthy sources and reliable guidance when they need it" (33), which is inducing an unprecedented need for responsible silence too.

Inadequacy of Regulatory Frameworks and Norms

The pandemic creates a generalized condition of resource-limited settings (RLSs), i.e., environments lacking means, specific knowledge, specialized personnel, MDs, and drugs within inappropriate medical locations. While this condition was already familiar to low- and middle-income countries, COVID-19 has overwhelmingly created RLS conditions in high-income countries, such as Europe, the USA, and Japan, for the first time since World War II. This demonstrates how regulatory frameworks for MDs and PPE are inadequate to RLS conditions. In fact, these regulations usually take into consideration standards that are too stringent and generic, proving impossible to fulfil in RLSs and, in times of the pandemic, difficult to adhere to universally. For example, the numerous tests and verifications required to assess the conformity of market respiratory protective equipment or eye protection equipment for healthcare purposes wasted time. One reason for this is that these standards are influenced by big manufacturers interested in having the largest market share. As a result, commercial standards for PPEs require testing in conditions that are not relevant for hospital workers (e.g., high temperatures typical of the heavy metals industry). Hence, international standards and norms followed the principle of generalism, losing universality and creating unnecessary burdens for small manufacturers (34). In this regard, the WHO has published, for the first time, technical guidance on PPE specifically relevant for healthcare settings (35, 36). Differing from ISO standards for masks and respirators, the WHO guidance focuses on essential parameters, such as filtering capability, fit, and breathability of masks.

Recalling the two aforementioned ethical perspectives, considering existing medical device regulation too generic to be universal does not mean adhering to the utilitarian-pragmatic current tout court. There may be a contextualized response, regulated on the basis of tests, complying with flexible standards, or rather standards that are purposely designed to take into account different niche conditions. However, the use of any object must be certified and not random, and subject to tests relating to its specific intended use. Only in this way can people and their rights be safeguarded, and can science prove to think deeply, act consciously, and remain silent when appropriate.

Contextualism is the basis of situational ethics (37), which seems to be the most adequate response to the specific needs of everyone while being able to face emergencies. In fact, it starts from the particular situation and tries to find universalizable answers, applying a heuristic and inductive method, progressive negotiations, and interdisciplinary exchanges.

Conclusions

One year after the start of the pandemic, the need for ethics guidance is still tangible in everyday circumstances and essential during crisis or in RLSs. Respecting fundamental ethical principles while negotiating among different criteria (hospitalization demands vs available ICUs, generalism vs particularism, action vs responsible-action) requires clear guidance, deep knowledge, and peer-to-peer discussion among experts of different disciplines. The need for extreme specialization should never result in the fragmentation of knowledge.

Exactly a century, i.e., the *Short Twentieth Century*, separates COVID-19 from the last pandemic, the so-called "Spanish Flu", which flagellated Europe in 1918-19. According to Hobsbawm, "*no period in history has been more penetrated by and more dependent on the natural sciences*" and "*yet no period, since Galileo's recantation, has been less at ease with it*". This chasm between scientists and the general public is still open and, in some cases, fomented by populisms, which leverage on people's fears evoking war atmospheres, which have nothing to do with the catastrophic failure of many national healthcare systems' response to this crisis. After a century, the dependence of medicine on biomedical science and engineering is evident, while their contribution to the definition of effective policies and norms is still negligible. Finally, the Cartesian fragmentation of knowledge, or rather "thinking in silos", has persisted across the last century, calling for the creation of more fora where multidisciplinary discussions can be promoted. Three main needs emerged clearly: the need for responsible thinking, the need for responsible action, and the need for responsible silence, when required and appropriate (38).

Abbreviations

WHO – World Health Organization; BME – biomedical engineering; EAMBES – European Alliance for Medical and Biological Engineering and Science; IUPESM – International Union for Physical and Engineering Sciences in Medicine; ICU – intensive care unit; SIAARTI – Società Italiana di Anestesia Analgesia Rianimazione e Terapia Intensiva (Italian Society for Anesthesia Analgesia Resuscitation and Intensive Care); NHS – National Health Service; MD – Medical device; PPE – personal protective equipment; DIY – do-it-yourself; CE – Conformité Européenne (European Conformity); RLS – resource-limited setting.

Notes

1 A. Maccaro, D. Piaggio, C. Dodaro, L. Pecchia, *Biomedical engineering and ethics: reflections on medical devices and PPE at the time of Covid-19*, in «BMC Medical Ethics», (2021) 22:130 https://doi.org/10.1186/s12910-021-00697-1, https://rdcu.be/cyAIH
2 www.ft.com/content/491a4af4-66e7-11ea-a3c9-1fe6fedcca75
3 www.nytimes.com/2020/03/20/us/politics/trump-coronavirus-supplies.html
4 www.lavanguardia.com/vida/20200411/48413026980/seat-detiene-produccion-respidradores-emergencia-descongestion-uci.html
5 https://uk-mobile-reuters-com.cdn.ampproject.org/c/s/uk.mobile.reuters.com/article/amp/idUKKCN21U0UI
6 www.franceinter.fr/coronavirus-8-500-respirateurs-produits-pour-rien
7 www.bbc.co.uk/news/health-52145140
8 www.telegraph.co.uk/news/2020/04/08/exclusive-three-nurses-forced-wear-bin-bags-ppe-shortage-test/

References

(1) French S. Between factualism and substantialism: Structuralism as a third way. *International Journal of Philosophical Studies.* 2018;26(5): 701–721.
(2) Morin E. From the concept of system to the paradigm of complexity. *Journal of Social and Evolutionary Systems.* 1992;15(4): 371–385.

(3) The Nuffield Council of Bioethics. Responding to the COVID-19 pandemic: Ethical considerations. 2020. Available from: www.nuffieldbioethics.org/news/responding-to-the-covid-19-pandemic-ethical-considerations. Accessed 19/6/2020.
(4) Department for Health of Ireland. Ethical framework for decision-making in a pandemic. 2020. Available from: https://assets.gov.ie/72072/989943ddd0774e7aa1c01cc9d428b159.pdf. Accessed 19/6/2020.
(5) Deutscher Ethikrat. Solidarity and responsibility during the Coronavirus crisis. 2020. Available from: www.ethikrat.org/en/press-releases/2020/solidarity-and-responsibility-during-the-coronavirus-crisis/. Accessed 19/6/2020.
(6) Comité De Bioética De España. Informe Del ComitComité De BioComitética De Espaé De Bioética De España Sobre Los Aspectos Bioéticos De La Priorización De Recursos Sanitarios En El Contexto De La Crisis Del Coronavirus. 2020. Available from: http://assets.comitedebioetica.es/files/documentacion/Informe CBE-Priorizacion de recursos sanitarios-coronavirus CBE.pdf. Accessed 19/6/2020.
(7) UNESCO. Statement on COVID-19: Ethical considerations from a global perspective. 2020. Available from: https://unesdoc.unesco.org/ark:/48223/pf0000373115. Accessed 19/6/2020.
(8) WHO. Ethical standards for research during public health emergencies: Distilling existing guidance to support COVID-19 R&D. 2020. Available from: www.who.int/blueprint/priority-diseases/key-action/liverecovery-save-of-ethical-standards-for-research-during-public-health-emergencies.pdf?ua=1. Accessed 19/6/2020.
(9) UNESCO. Ethics in research in times of pandemic COVID-19. 2020. Available from: https://en.unesco.org/news/ethics-research-times-pandemic-covid-19. Accessed 19/6/2020.
(10) Meyfroidt G, Vlieghe E, Biston P, De Decker K, Wittebole X, Collin V, et al. Ethical principles concerning proportionality of critical care during the COVID-19 pandemic: Advice by the Belgian Society of IC medicine. 2020. Retrieved 2/4/2020, from: www.hartcentrumhasselt.be/professioneel
(11) Berlinger N, Wynia M, Powell T, Hester DM., Milliken A, Fabi R, et al. Ethical framework for health care institutions responding to novel Coronavirus SARS-CoV-2 (COVID-19) guidelines for institutional ethics services responding to COVID-19. *Safeguarding Communities, Guiding Practice.* 2020;12(3), 1–12.
(12) Smith MJ, Ahmad A, Arawi T, Dawson A, Emanuel EJ, Garani-Papadatos T, et al. Top five ethical lessons of COVID-19 that the world must learn. *Wellcome Open Research.* 2021;6.
(13) Smith MJ, Upshur RE. Learning lessons from COVID-19 requires recognizing moral failures. *Journal of Bioethical Inquiry.* 2020;17(4):563–566.
(14) Smith MJ, Upshur RE, Emanuel EJ. Publication ethics during public health emergencies such as the COVID-19 pandemic. *American Journal of Public Health.* 2020;110(7):947–948.
(15) Emanuel EJ, Persad G, Upshur R, Thome B, Parker M, Glickman, Zhang C, Boyle C, Smith M, Phillips JP. Fair allocation of scarce medical resources in the time of Covid-19. New England Journal of Medicine. 2020 May 21;382(21):2049–55.
(16) WHO. Resources on ethics and COVID-19. 2020. Available from: www.who.int/ethics/topics/outbreaks-emergencies/covid-19/en/. Accessed 19/6/2020.
(17) WHO. Addressing human rights as key to the COVID-19 response. 2020. Available from: www.who.int/publications-detail/addressing-human-rights-as-key-to-the-covid-19-response. Accessed 19/6/2020.

(18) Comitato nazionale per la bioetica. Covid 19: Clinical decision-making in conditions of resource shortage and the "pandemic emergency triage criterion". 2020. Available from: http://bioetica.governo.it/media/3987/p136_2020_covid-19-la-decisione-clinica-in-condizioni-di-carenza-di-risorse-e-il-criterio-del-triage-in-emergenza-pandemica.pdf. Accessed 19/6/2020.
(19) WHO. Preparedness. 2020. Available from: www.who.int/environmental_health_emergencies/preparedness/en/. Accessed 19/6/2020.
(20) Habermas J. The only cure is solidarity from www.classlifestyle.com/news/42575/kura-e-vetme-eshte-solidariteti-nga-jrgen-habermas/eng. 2020. Accessed 19/6/2020.
(21) Vergano M, Bertolini G, Giannini A, Gristina GR, Livigni S, Mistraletti G, et al. *Clinical ethics recommendations for the allocation of intensive care treatments in exceptional, resource-limited circumstances: The Italian perspective during the COVID-19 epidemic*: Springer; 2020.
(22) Rosenbaum L. Facing Covid-19 in Italy – Ethics, logistics, and therapeutics on the epidemic's front line. *New England Journal of Medicine*. 2020;382(20):1873–1875.
(23) Nicoli F, Gasparetto A. Italy in a time of emergency and scarce resources: The need for embedding ethical reflection in social and clinical settings. *Journal of Clinical Ethics*. 2020;31(1):92–94.
(24) Ranney ML, Griffeth V, Jha AK. Critical supply shortages – The need for ventilators and personal protective equipment during the Covid-19 pandemic. *New England Journal of Medicine*. 2020;382(18):e41.
(25) Sokolov P, Ivanova J. 17th century political Cartesianism and its opponents, or imaging the state from point fixe. *Higher School of Economics Research Paper No. BRP*, 8. 2012.
(26) Jones D. *Censorship: A world encyclopedia*: Routledge; 2001.
(27) Maasen S, Weingart P. *Democratization of expertise?: Exploring novel forms of scientific advice in political decision-making* (Vol. 24): Springer Science & Business Media; 2006.
(28) Bennett B. *Logically fallacious: The ultimate collection of over 300 logical fallacies* (Academic Edition): EBookIt. com; 2017.
(29) Regulation EU 2017/745 of the European Parliament and of the Council of 5 April 2017 on medical devices, amending Directive 2001/83/EC, Regulation (EC) No 178/2002 and Regulation (EC) No 1223/2009 and repealing Council Directives 90/385/EEC and 93/42/EEC (2017). In the European parliament and the council of the European Union (Ed.).
(30) Jonas H. *The imperative of responsibility: In search of an ethics for the technological age*. University of Chicago. 1984;202.
(31) Bentham J, Mill JS. *Utilitarianism and other essays*: Penguin UK; 2004.
(32) Dewey J. *The essential Dewey: Pragmatism, education, democracy* (Vol. 1): Indiana University Press; 1998.
(33) WHO. Novel Coronavirus (2019-nCoV) situation report – 13. 2020.
(34) Pecchia L, Piaggio D, Maccaro A, Formisano C, Iadanza E. The inadequacy of regulatory frameworks in time of crisis and in low-resource settings: Personal protective equipment and COVID-19. *Health and Technology*. 2020 Nov;10(6):1375–83.
(35) WHO. Rational use of personal protective equipment for COVID-19 and considerations during severe shortages: Interim guidance, 23 December 2020. (No. WHO/2019-nCoV/IPC_PPE_use/2020.4). World Health Organization.

(36) WHO. Technical specifications of personal protective equipment for COVID-19: *interim guidance, 13 November 2020* (No. WHO/2019-nCoV/PPE_specifications/2020.1). World Health Organization..
(37) Fletcher JF. *Situation ethics: The new morality*: Westminster John Knox Press; 1997.
(38) Maccaro A, Piaggio D, Pagliara S, Pecchia, L. The role of ethics in science: a systematic literature review from the first wave of COVID-19. Health and Technology. 2021 Sep;11(5):1063-71.

The Role of Ethics in Science: A Systematic Literature Review from the First Wave of COVID-19[1]

Background

The spread of COVID-19 begot what the World Health Organization (WHO) defined a "pandemic" (1), an emergency condition that has often been compared to other dramatic events in history (e.g. the Spanish flu in 1918). However, on closer inspection, COVID-19 can be regarded as an unprecedented event with its own specificity.

During the first months of the pandemic, the numerous emerging issues were deeply interwoven with ethics, as it is unanimously recognized by the works of politics, medicine, and science. The interdisciplinarity of such works and their continuous and necessary reference to ethics reinforce the belief that bioethics, i.e., a "bridge" between different fields of knowledge, looks more and more "towards the future" (2). In fact, in the first few months, the contributions of National Ethics Committees, International Organizations, National Bodies and Professional Association to this topic (3) were numerous, very rapid, and accompanied a very heated debate on ethical implications between the population and experts from various backgrounds. Nonetheless, the continuous reference to the word "ethics" in the first publications after the declaration of COVID-19 pandemic (4–7) highlighted the need to clarify the most relevant ethical problems related to the scientific community. Hence, it was decided to focus on and deepen the subjects that aroused the interest of specialists the most. Specifically, existing works of reconstruction and/or comparison among national and international documents on the relationship between ethics and COVID-19 were not taken into consideration, to avoid delving into or discoursing about already exhaustively discussed topics.

Consequently, a systematic literature review was opted for. Such review was conducted by an interdisciplinary working group, comprising biomedical engineers and bioethicists, following a multidisciplinary approach, which overcomes the outdated Cartesian model of the separation of knowledge into "silos" of disciplines (8, 9). To facilitate the screening process, the papers were grouped by macro-areas that were pinpointed through a thematic analysis (10), as explained

DOI: 10.1201/9781003436348-4

in the methods. In particular, it was decided to analyze only the themes related to policy, technology, resource allocation, and low- and middle-income countries (LMICs) as they fall within our competences and seem to underline relevant ethical questions which are not always adequately or exhaustively discussed. However, the theme of resource allocation was excluded *a priori*, because it is a trite topic, currently "abused", on which anyone expresses their opinion independently from their level of expertise and knowledge. Moreover, many papers, which had a reference to ethics in the title, but which, in fact, did not deal with analyzing the ethical implications of the investigated issues, were excluded. In fact, the ultimate purpose of this work was to understand the role that scientists recognize in ethics and the problems related to it: is it seen as a "humanitarian" addition to technical issues or as a structural element and perspective from which to analyze specialistic issues?

Therefore, this paper reports a collection of the most relevant and less conventional ethics challenges related to COVID-19 published in peer-reviewed journals indexed in PubMed, and analyzed through our multidisciplinary lens. These ethical issues, which emerged during the first wave of the pandemic, were then rediscussed a year later in order to assess whether the first bioethical perspectives related to COVID-19 were biased by the close succession of the events being analyzed, or whether they were detached enough and the raised issues remain current a year on.

Methods

Systematic Literature Review

Given the large number of papers and documents that have been and are currently being published since the start of the pandemic, we decided to conduct a systematic literature review. PubMed was selected as the only database to identify all the contributions published from 01/01/2020 up to 19/05/2020 responding to the topics of COVID-19 and ethics. The search string was constructed with the following terms combined with the Boolean operators AND/OR: "ethics", "ethical", "bioethics", "COVID", "sars cov 2", "coronavirus" ((ethics OR ethical OR bioethics) AND (COVID OR sars cov 2 OR coronavirus)). We judged eligible only the papers with full text available and in English. In order to facilitate the clustering and further screening of the retrieved articles, they were divided into the most recurring themes and per geographical area. To this regard, the countries were grouped according to the following macro-groups, based on the retrieved papers: North America, Europe, LMICs (e.g., Tunisia), Asia, and Israel. Particular attention was dedicated to the contributions regarding LMICs. Finally, only the paper pertaining to our field of expertise, i.e., technologies and policy, made it through the final selection. During this process, two authors independently screened all the titles and abstracts for eligibility. Full texts were considered if the selection was unclear. A third author reviewed and checked the results of the screening search. Any discrepancy was resolved by discussion among all the authors. All the relevant papers were analyzed, summarized, and coded to facilitate the reading.

Results

Systematic Literature Review

Figure 1.1 shows the electronic database search and the selection process. Only 38 papers out of the initial 233 were eligible to be included in our study, which focuses on technology and policy. Figure 1.2 shows the division in macro-groups and

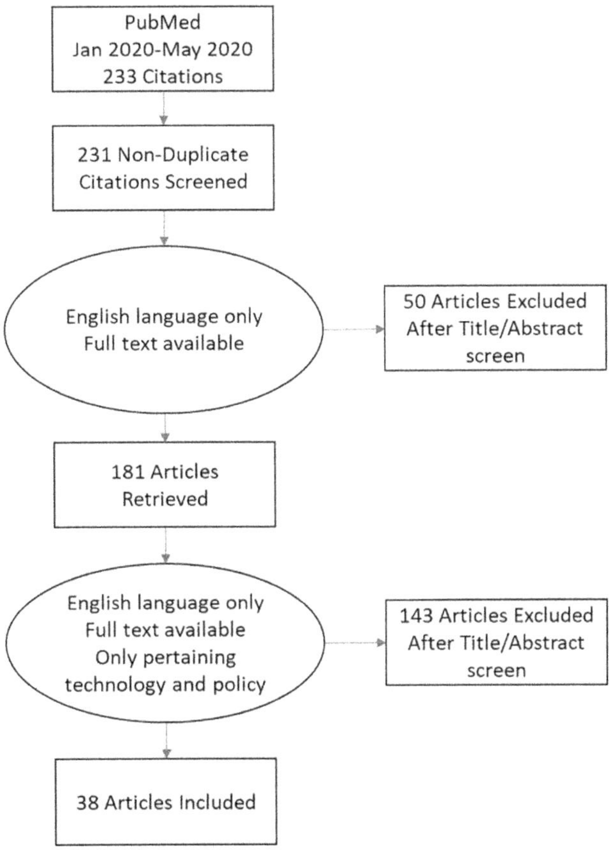

FIGURE 1.1
The flowchart of the systematic literature review.

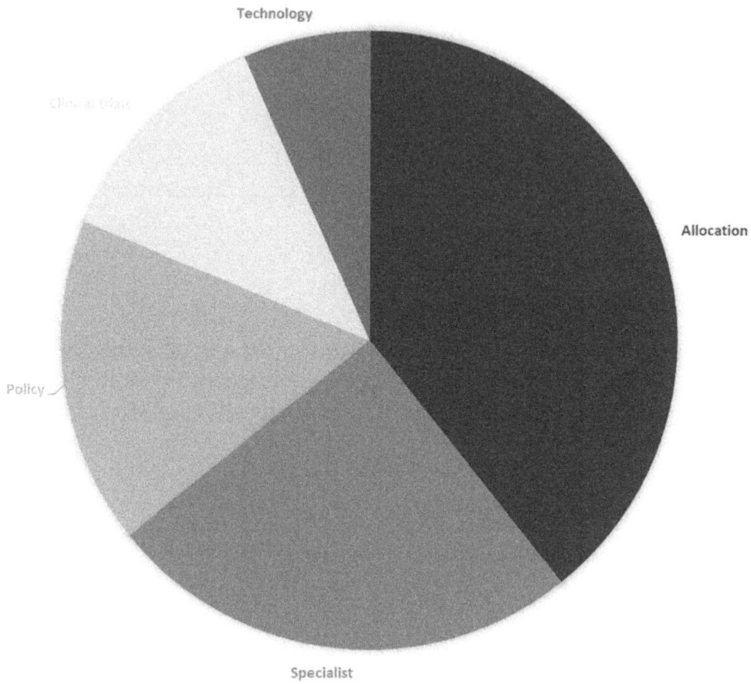

FIGURE 1.2
The distribution of the articles in five macro-areas.

Figure 1.3 the division per geographical area, when applicable. Online Resource 1 presents all the selected papers organized by recurring themes.

The five pinpointed macro-groups of specific works on ethics and COVID-19 were: Policy, Resource allocation, Specialists, Clinical trials, and Technology. The theme of policies, i.e., of the public responses to the crisis, is the one that piques scientists' interest the most, according to our review. In this regard, as already mentioned, our review is not exhaustive and does not aim at including all the political guidelines of the various countries, because its hermeneutical horizon (11) is the point of view of science. Notwithstanding, it was possible to identify a series of specific works on different countries, which allowed comparisons between different areas of the world: LMICs, Asia, Europe, and North America.

The theme of resource allocation is also extremely well-liked. In fact, the first wave of the pandemic caused a scarcity of resources, globally, without any distinction: from personal protective equipment (PPE) to medical devices (MDs) (e.g., ventilators and respirators), beds, drugs for COVID-19 patients or patients suffering from other pathologies, health personnel, and COVID-19 tests. Most specifically, this situation of emergency abruptly showed the lack of competent ad-hoc bodies. This begot extremely heterogeneous approaches to ethics by different individuals, regardless of their competence in ethics.

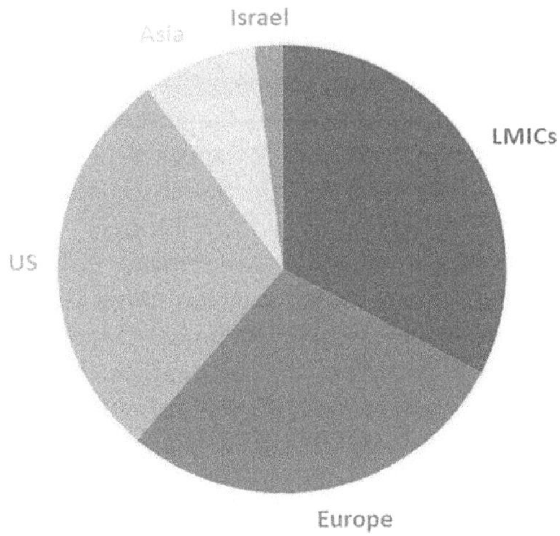

FIGURE 1.3
The distribution of the articles per geographical area.

Among the retrieved publications, there were also numerous articles on ethics and COVID-19 written by specialists from various medical fields (e.g., geriatrics, psychiatry, surgery, oncology, and dentistry). Other publications focuses on all healthcare workers (HCWs), such as nurses and midwives, who experienced profound difficulties in this particular historical moment, the common denominator being allocation problems.

A significant number of works were related to clinical trials, in particular concerning vaccines and drugs, dealing with detecting the widespread condition of researchers who had to readdress the scopes of their research, always respecting high ethical standards, and safeguarding the testers.

Finally, the last area is that of technology for COVID-19 (e.g., e-Health) and all the related ethical issues.

Discussions

The Duality of Trust: On the Trustworthiness of Governments and Public Trust

Most of the articles included the topic of *infodemic*, i.e., "an over-abundance of information – some accurate and some not – that makes it hard for people to find trustworthy sources and reliable guidance when they need it" (12). Misinformation circulating through global digital social networks in the first months of this pandemic was focused on trust in governments and policy makers (13–15), questioning the principles of legitimacy, and responsibility related to information verification and sharing.

But who was the source of such disinformation?

Larson HJ et al. (15) claimed that it is was the governments who repressed information hoping to calm anxious publics, as it happened in Iran (15, 16) or in China, where healthcare workers, who told the truth about the spread of the virus and the scarcity of resources, were looked at as "whistle-blowers" and forced to withdraw their declarations (17). Moreover, some governments deliberately released supposedly reassuring misinformation, risking undermining their own credibility and their abilities to help people counter real health threats. According to Limaye RJ et al. (13), there is a need for joint action between government agencies and social media companies for fact-checking and even removal of false or outdated information. On the other hand, according to Bastani P (16), there is the need of an active and effective presence of health professionals and authorities on social media, due to the poor legal supervision of online content.

Low-quality research contributed to this misinformation too. In fact, the high demand for information caused an acceleration in reporting scientific results, with many journals publishing without any peer-review and offering open-access to everyone. A proxy for the high number of low-quality research on COVID-19 is also the unprecedented high number of retracted papers. To this regard, we searched for papers on COVID-19 or SARS-Cov-2 and the previous epidemics/pandemics (i.e., Avian Flu, Swine Flu, MERS) both on the Retraction Watch Database and OvidSP. As a result, as of May 2021, 124 out of 264530 papers on COVID-19 (4.68 retractions per 10,000) were retracted, compared to 1.16 per 10,000 papers concerning the previous pandemics/epidemics.

Apart from being unethical, reporting poor quality outcomes is a kind of research misconduct (18) and inaccuracy. This could also have other consequences, for example exacerbating stigma and discrimination against particular populations (19). In fact, Chowkwanyun et al. (20), analyzing the information circulated in the media in Wisconsin and Michigan on the high percentage of black people affected by Coronavirus, argued that there is always the need for contextualization unless we want to foster harmful myths and misunderstanding which undermine the goal of eliminating health inequities. The fear of stigmatization towards specific groups of people is likely to worsen if they are individuals. In this case, the ethical dilemma is the balance between personal and collective interests.

In hindsight, the public health measures that have been implemented can be of different types. Sulmasy et al. (21) identifies four of them:

1. Contact tracing through the self-reporting of recent close interactions by people known to be infected with COVID-19. In this regard, Luo H et al. (22) reported the example of an online questionnaire that circulated in China as an internet approach in COVID-19 participatory surveillance.
2. More "draconian" health measures: new surveillance technologies that employ facial recognition, security cameras, and phone GPS monitoring could

attempt to identify everyone who spent at least fifteen minutes within six feet of every infected individual. Each contact could be forcibly quarantined. In this regard, Ferretti et al. (23) underlined that apps could also be used as a preventive approach. However, as Vokinger et al. (24) affirmed, trustworthiness and integrity of contact tracing apps should be assessed with a framework. A review of the tracing apps for the management of COVID-19 can be found in (25), where the authors underlined that among the advantages there were the increase of personal freedom, of personal feelings of safety, and the improvement of the management of the quarantine. However, governments should implement policies to outline requirements for these apps and should safeguard privacy, access, transparency, the protection and use of these data (24, 25). In particular, Santow et al. (26) underlined the need for a legal framework to regulate artificial intelligence and data sharing, as they can be the cause of discrimination and violation of human rights.

3. Voluntary contact tracing: it relies on self-led contact tracing. However, it is not perfect because patients might not remember all their recent contacts.
4. The public naming of infected individuals: in this regard, Sulmasy DP et al. (21) described the case of the Prestigious University, in which the communication via email about the testing positive of a staff member raised a heated debate regarding the request to reveal the identity of the subject in order to maximize public health benefit and slow the spread of the virus. One of the authors argued that this could breach confidentiality and be harmful to the patient's privacy, who is free to decide whether to make a voluntary disclosure. However, another author believed that the confidentiality breach is morally mandatory to decrease the risk of contagion for other members of the University, because the duty towards the community has priority over the right of confidentiality. This is also discussed by Persad et al. (27), who reported the proposal of some states (i.e., Chile, Germany, and the UK) to implement "certifications of immunity" or "immunity passports" for those who have had COVID-19 or who will have received the vaccine in the future. In line with the principle of the "least restrictive alternative" to achieve public health objectives, the author believe that this tracking measure is not unethical and cannot be compared to the yellow star that the Nazis forced the Jews to wear, because it is not a form of discrimination. However, it needs careful implementation and scientific support to be ethical in practice.

Another category of individuals who place the interests of society before their own is that of frontline healthcare workers (HCWs) who are forced to work in precarious conditions at their personal risk (28). However, for them there is also another ethical dilemma: that of protecting, together with one's own health, also one's family and loved ones, which often clashes with the duty to treat patients, sometimes in the absence of PPE (29, 30). According to McConnell D. et al. (31), there are several factors that adjust the burden of protecting one's family. The authors argued that the moral demands of "Samaritanism" (i.e., one should go out of one's way to help someone else if it entails a little cost to oneself) do not imply that HCWs take on the risks and burdens associated with treating COVID-19 to save several lives. Likewise, Thomas JP (30) denounced the low quantity and quality of PPE and the inadequacy of the related guidance issued by Public Health England. Moreover, he appealed to the precautionary principle and praised the ethical framework of Beauchamp and Childress (32) that encourages to counterbalance beneficence with non-maleficence. Always according to Thomas JP (30), political leaders have the moral duty to be open and honest, when informing all frontline HCWs of their own personal risks in caring for COVID-19 patients.

Nonetheless, as in all the aspects of life, economical evaluations cannot be excluded. Hilsenrath PE et al. (33) presented a very sensitive issue, that of confronting the medical duty to save lives and the reasoning of economists, who invite people to make decisions on appropriate costs. The author lucidly underlined that although the issue is often bypassed by political leaders, especially Americans, it should be faced by considering an ethical balance, certainly painful, between the ethical and economic damage that countries are experiencing in this historical moment.

Overall, what is required from governments is trust. People place their trust, their lives, health, and economic situation in the sensible and transparent decision-making of governments. Therefore, as Thomas JP claimed, "it is a reasonable expectation to hold our modern governments to the corresponding standards of our modern health professionals: specifically, transparency in decision-making and the duty of candidate" (30). Similarly, Lewnard JA et al. (34) sustained that policymakers maintain the public's trust through the use of evidence-based interventions and fully transparent fact-based communication. To give an example of this, the author also reported on mathematical modelling of the viral transmission under different scenarios to generate evidence of the efficacy of social distancing interventions.

On the other hand, Chaari L. et al. (35), supporter of 3P (predict, prevent, and personalize) medicine, proposed "real-time" monitoring based on randomized laboratory tests as a source of evidence. However, this clashed with the problem of the lack of tests and the discrepancy between officially recorded and real infection cases. In fact, this could lead to incorrect political decisions heavily influencing the future of a country (e.g., a long-term economic recession due to over-protection of the population, or a post-containment pandemic rebound due to an under-protection of the population). Furthermore, all the measures required from governments (i.e., testing, screening, contact tracing, social distancing, and travel restriction) must be as inclusive as possible in particular with vulnerable communities (i.e., homeless, those without insurance or employment, disabled, immigrant, and prisoners) (14, 34). This also means being aware that the use of technology to combat the spread of COVID-19 could exacerbate racial, socioeconomic, and geographic disparities for populations that lack access to reliable internet access, devices or technological literacy (36). Moreover, during the pandemic, the benefits of technologies that allow relationships beyond social distancing have increasingly been experienced and could be a valid assistance tool for the elderly (37) and favor the communication of hospitalized patients with their families (36). This response should also involve millennials and

Generation Z more (38), as, despite the negative stereotypes that circulate around them, they could offer valuable help to overcome this crisis.

Certainly, the use of technology, if regulated, can be of help to scientists. Indeed, O'Reilly-Shah et al. (39) highlighted the shortcomings in the US healthcare IT infrastructures, underlining the importance of the interoperability of healthcare data, which should refrain from the proprietary control of vendors and be accessible to healthcare providers, especially in times of crisis.

More generally, it could be said that technology and data sharing are particularly important for disseminating knowledge worldwide. Momtazmanesh et al. (40) sustained that sharing and solidarity are at the base of an indispensable international collaboration to fight the current and future pandemics. The pandemic has increasingly shown the need for an international ethical-political coordination framework (23), aiming at reducing disparities. In this regard, under the category of *policy*, many contributions regarded LMICs. One of the recurrent themes was the stress on already overburdened and underfunded public healthcare systems in Africa, India, and Latin America (41–45). In a more general comparison between the North and the South of the world, Schuklenk et al. (43) affirmed that there is a significant difference in the number of available intensive care unit beds per population and that the access to cures is often wealth-based, as many hospitals in the South are private. However, Krishna et al. (42) stated that COVID-19 is just another drop in an already full vase, as the Indian healthcare system is already plagued by internal issues (e.g., no access to medicines, vested political interests etc.). All in all, the policies against COVID-19 are, once again, putting the poorest countries and people at risk (43, 46). In India, for example, the poor were confined in ghettos without a proper social security net and their conditions were exacerbated by the lockdown (46). Other approaches used in LMICs to fight COVID-19 include the use of fear appeals to regulate people's behaviors (47). However, it can be argued whether this technique is ethical and acceptable or not. On the other hand, Kapata et al. (48) positively affirmed that Africa was readier than for the previous epidemic outbreaks.

Other diseases, such as vector-borne and non-communicable diseases, and community-acquired infections are often included among other stressors (41, 42). Krishna et al. (42) also affirmed that notwithstanding this, COVID-19 obtained much more funding compared to other existing deadly diseases (e.g., diarrhea).

Another element contributing to exacerbating the situation is the lack of ethics committees in the hospitals to regulate and ease the work of healthcare operators. In fact, the latter do not only have to face very difficult choices in an environment dominated by material scarcity and public distrust (41), but they also have to work in precarious conditions and their striking for this matter could be seen as patient abandonment (43). With regard to this, Gopichandran et al. (44) recalled the *reciprocity principle*, according to which the state should protect the interests of the healthcare workers, who risk their lives to care for those who are infected.

Some authors (16, 44, 45, 49) agreed on the need for equity and clarity in the way governments inform citizens, so as to not undermine their trust. Most importantly, it was unanimously stated that policies for COVID-19 should always be adapted to different contexts, above all the ones related to minorities, in the respect of traditional beliefs (50), and to avoid exacerbations of pre-existing gaps between the rich and the poor (46). To this purpose, self-determination is key (51): in fact, LMICs should independently shape their response, relying on international partners in a critical manner. For instance, LMICs should avoid the unconditional acceptance of measures adopted in high-income countries that would be inappropriate in resource limited settings (52–54).

Overall, the number and significance of the ethical problems that emerged which makes us understand how ethics should be increasingly involved in guiding political and scientific legal reasoning, lead us to disagree with Stoeklé et al. (55), who, by separating political discussions, scientific knowledge, and ethics writes: "now really isn't the time for ethical reflections" and "ethics is only really useful if you have the time, and right now, time is exactly what we do not have."

Conclusions

This paper presented a systematic overview of scientific papers investigating ethical issues during the first wave of this pandemic. The papers highlighted some recurrent themes, namely the allocation of scarce resources, infodemic, HCWs' duty to treat versus personal protection, privacy, and the safeguarding of minorities.

Reviewing such themes, a year after the outbreak of COVID-19, highlights several facts.

The infodemic has been as devastating as the virus. While it could have been acceptable in the first months of this pandemic, it is somehow surprising that after three waves of COVID-19, there is still confusion around trustworthy sources and reliable guidance. A year after, there is still misinformation spread through digital social networks, although the object of the discussion has changed, focusing now on vaccination safety and scarcity, rather than MDs and PPE. The same doubts raised by Larson (15), Limaye (13) and Bastani (16) in March 2020 on the intentional repression of information perpetuated by governments for calming anxious publics during the first wave, have been recently repeated in regard to the surge of COVID-19 that affected India since March 2021.

After more than 12 months from the first wave, the ethical concerns on the stigma and discrimination against particular populations is still relevant in many ways. For instance, several authors hypothesize a causal link between COVID-19 and the surge of violent acts towards Asians living in the USA (56, 57). On the global scale, this discrimination will certainly not be mitigated by the delay in vaccinating populations in LMICs.

Moreover, the ethical concerns arising from balancing the need for track and trace with the risks for privacy, seem to be still unresolved. After a year and a half into the pandemic, the same skepticism has crossed many COVID-19 waves resulting in several Apps being abandoned for more traditional phone-based methods. Similarly, the envisioned "real-time" monitoring based on randomized laboratory tests is still far from being possible and could still lead to incorrect political decisions heavily influencing the future of a country. One year on, in fact, although this risk has lowered globally, it still is critical for some LMICs, where several challenges hinder COVID-19 testing (58).

Concerns on immunity certifications and passes, are yet far from being resolved, although the attention has now shifted towards the vaccines.

Concerning HCWs, the memory of recognizing their efforts by cheering and clapping from the windows is far.

The scarcity of PPE and MDs seems to be overcome now, in many high-income countries, also thanks to the massive effort of United Nation agencies lead by the World Health Organization. Yet, the problem is far from being solved without structural interventions, as demonstrated by the surge in cases in India.

Regarding the ethical issues on the appropriateness of economical evaluations and the transparency of political decisions, raised by authors during the first months of this pandemic, there is a need for more time before one can make unbiased reflections.

Inclusiveness seems to be one of the forgotten principles during this pandemic. While this issue was raised from the very start, the question remains open for several groups of populations, which have been not prioritized during the first wave, such as children, poor, unstable workers, chronic patients, and single parents.

Overall, this review identified a number of papers referring to ethics in their title, which in the end presented little or no contribution to the ongoing discussion on ethical issues arising from COVID-19. While the attention of many scholars to this important topic is remarkable, we conclude this review with the doubt that ethics is still considered as a tick-box exercise by many, when not a "humanitarian embellishment" of their technical work.

Conversely, ethics should be deeply interwoven with science, theoretically and methodologically. Besides helping the evaluation of the arising issues, ethics could help scientists ponder on the risk-benefit balance of their publications, and also on the final purpose of their work, i.e., the progress of humanity. The unprecedented number of retracted papers suggests that during this pandemic the race to participate in the infodemic overcame the sense of responsibility, which should have imposed, in many cases, a responsible silence (59).

List of Abbreviations

WHO – World Health Organization; LMICs – low- and middle-income countries; PPE – personal protective equipment; MDs – medical devices; HCWs – healthcare workers.

Note

1 A. Maccaro, D. Piaggio, S. Pagliara, L. Pecchia, *The role of ethics in science: a systematic literature review from the first wave of COVID-19*, in «Health and Technology» (Springer), 03/06/2021: https://link.springer.com/content/pdf/10.1007/s12553-021-00570-6.pdf

References

(1) Cucinotta D, Vanelli M. WHO declares COVID-19 a pandemic. *Acta Bio Medica: Atenei Parmensis*. 2020;91(1):157.
(2) Bingham N. Potter, Van Rensselaer. Bioethics: Bridge to the future. Englewood Cliffs, NJ Prentice-Hall, 1971. *Science Education*. 1972;56(3):440–441.
(3) WHO. Resources on ethics and COVID-19. 2020 [cited 5/6/2020]; Available from: www.who.int/ethics/topics/outbreaks-emergencies/covid-19/en/
(4) Rosenbaum L. Facing Covid-19 in Italy – Ethics, logistics, and therapeutics on the epidemic's front line. *New England Journal of Medicine*. 2020;382(20):1873–1875.
(5) WHO. International working group on Ethics and COVID-19. 2020 [cited 2/4/2021]; Available from: www.who.int/news/item/20-03-2020-who-has-established-an-international-working-group-on-ethics-covid-19
(6) Zohny H. Pandemic ethics: Covid-19 shows just how much of ethics depends on (good) data. 2020 [cited 2/4/2021]; Available from: https://blog.practicalethics.ox.ac.uk/2020/03/pandemic-ethics-covid-19-shows-just-how-much-of-ethics-depends-on-good-data/
(7) Klitzman R. How hospitals will decide who lives and who dies in the COVID-19 crisis. 2020 [cited 2/4/2021]; Available from: https://thehill.com/opinion/healthcare/490034-how-hospitals-will-decide-who-lives-and-who-dies-in-our-coronavirus-crisis
(8) Schnurman Z, et al. Knowledge silos: Assessing knowledge sharing between specialties through the vestibular schwannoma literature. *Journal of Neurosurgery*. 2017;129(5):1278–1285.
(9) Morin E. From the concept of system to the paradigm of complexity. *Journal of Social and Evolutionary Systems*. 1992;15(4):371–385.
(10) Tuckett AG. Applying thematic analysis theory to practice: A researcher's experience. *Contemporary Nurse*. 2005;19(1–2):75–87.
(11) Gadamer H.-G. *Truth and method* (J Weinsheimer and DG Marshall, trans.): Continuum; 1989.
(12) WHO. Technical focus: Risk communication and community engagement. 2020 [cited 5/6/2020]; Available from: www.who.int/docs/default-source/coronaviruse/situation-reports/20200202-sitrep-13-ncov-v3.pdf?sfvrsn=195f4010_6
(13) Limaye RJ, et al. Building trust while influencing online COVID-19 content in the social media world. *The Lancet Digital Health*. 2020;2(6):e277–e278.
(14) Berger ZD, Evans NG, Phelan AL, Silverman RD. Covid-19: control measures must be equitable and inclusive. Bmj. 2020 Mar 20;368.

(15) Larson HJ, Blocking information on COVID-19 can fuel the spread of misinformation. *Nature*. 2020;580(7803):306.
(16) Bastani P, Bahrami MA. COVID-19 related misinformation on social media: A qualitative study from Iran. *Journal of Medical Internet Research*. 2020 Apr 5.
(17) Zhu J. Is it ethical to be a "whistleblower" during COVID-19 pandemic? – Ethical challenges confronted by health care workers in China. *Journal of Advanced Nursing*. 2020 Aug;76(8):1873.
(18) Rapid outbreak response requires trust. *Nature Microbiology*. 2020;5(2):227–228.
(19) Smith MJ, Upshur RE, Emanuel EJ. *Publication ethics during public health emergencies such as the COVID-19 pandemic*: American Public Health Association; 2020.
(20) Chowkwanyun, M, Reed AL Jr. Racial health disparities and Covid-19—Caution and context. *New England Journal of Medicine*. 2020 Jul 16;383(3):201–3.
(21) Sulmasy DP, Veatch RM. Should institutions disclose the names of employees with Covid-19? *Hastings Center Report*. 2020 May;50(3):25–7.
(22) Luo H, Lie Y, Prinzen FW. Surveillance of COVID-19 in the general population using an online questionnaire: Report from 18,161 respondents in China. *JMIR Public Health and Surveillance*. 2020;6(2):e18576.
(23) Ferretti L, Wymant C, Kendall M, Zhao L, Nurtay A, Abeler-Dörner L, Parker M, Bonsall D, Fraser C. Quantifying SARS-CoV-2 transmission suggests epidemic control with digital contact tracing. *Science*. 2020 May 8;368(6491):eabb6936.
(24) Vokinger KN, Nittas V, Witt CM, Fabrikant SI, Von Wyl V. Digital health and the COVID-19 epidemic: An assessment framework for apps from an epidemiological and legal perspective. *Swiss Medical Weekly*. 2020;150:w20282.
(25) Parker MJ, Fraser C, Abeler-Dörner L, Bonsall D. Ethics of instantaneous contact tracing using mobile phone apps in the control of the COVID-19 pandemic. *Journal of Medical Ethics*. 2020 Jul 1;46(7):427–31.
(26) Santow E. Emerging from AI utopia. Science. 2020 Apr 3;368(6486):9–9.
(27) Persad G, Emanuel EJ. The ethics of COVID-19 immunity-based licenses ("immunity passports"). *JAMA*. 2020 Jun 9;323(22):2241–2.
(28) Harkin DW. Covid-19: Balancing personal risk and professional duty. *BMJ*. 2020 Apr 29;369:m1606.
(29) Bakewell F, Pauls MA, Migneault D. Ethical considerations of the duty to care and physician safety in the COVID-19 pandemic. *Canadian Journal of Emergency Medicine*. 2020 Jul;22(4):407–10.
(30) Thomas JP, Srinivasan A, Wickramarachchi CS, Dhesi PK, Hung YM, Kamath AV. Evaluating the national PPE guidance for NHS healthcare workers during the COVID-19 pandemic. *Clinical Medicine*. 2020 May 1;20(3):242–7.
(31) McConnell D. Balancing the duty to treat with the duty to family in the context of the COVID-19 pandemic. *Journal of Medical Ethics*. 2020;46(6):360–363.
(32) Beauchamp T , Childress J. *Principles of biomedical ethics: Marking its fortieth anniversary*: Taylor & Francis; 2019.
(33) Hilsenrath PE. Ethics and economic growth in the age of COVID 19: What is a just society to do? *Journal of Rural Health*. 2020 Jan 1;37(1).
(34) Lewnard JA, Lo NC. Scientific and ethical basis for social-distancing interventions against COVID-19. *The Lancet. Infectious Diseases*. 2020 Jun 1;20(6):631–3.
(35) Chaari L , Golubnitschaja O. Covid-19 pandemic by the "real-time" monitoring: The Tunisian case and lessons for global epidemics in the context of 3PM strategies. *EPMA Journal*. 2020 Jun;11(2):133–8.

(36) Hart JL, Turnbull AE, Oppenheim IM, Courtright KR. Family-centered care during the COVID-19 era. *Journal of Pain and Symptom Management*. 2020 Aug 1;60(2):e93–7.
(37) Meinert E, et al. Agile requirements engineering and software planning for a digital health platform to engage the effects of isolation caused by social distancing: Case study. *JMIR Public Health and Surveillance*. 2020;6(2):e19297.
(38) Gharzai LA, Beeler WH, Jagsi R. Playing into stereotypes: Engaging millennials and generation Z in the COVID-19 pandemic response. *Advances in Radiation Oncology*. Jul 1;5(4):679–81.
(39) O'Reilly-Shah VN, Gentry KR, Van Cleve W, Kendale SM, Jabaley CS, Long DR. The COVID-19 pandemic highlights shortcomings in us health care informatics infrastructure: A call to action. *Anesthesia and Analgesia*. 2020 Aug 1;131(2):340–4.
(40) Momtazmanesh S, Ochs HD, Uddin LQ, Perc M, Routes JM, Vieira DN, Al-Herz W, Baris S, Prando C, Rosivall L, Latiff AH. All together to fight novel coronavirus disease (COVID-19). *The American Journal of Tropical Medicine and Hygiene*. 2020 Jun;102(6):1181.
(41) Litewka SG, Heitman E. Latin American healthcare systems in times of pandemic. *Developing World Bioethics*. 2020 Jun;20(2):69–73.
(42) Krishna S. The Covid-19 panic and the global North-South divide – A perspective. *Indian Journal of Medical Ethics*. 2020 Apr 1(2):167–8.
(43) Schuklenk U. The ethical challenges of the SARS-CoV-2 pandemic in the global south and the global north – Same and different. *Developing World Bioethics*. 2020 Jun;20(2):62.
(44) Gopichandran V, Subramaniam SU. Response to Covid-19: An ethical imperative to build a resilient health system in India. *Indian Journal of Medical Ethics*. 2020 Apr 1(2):1–4.
(45) Garg S, Bhatnagar N, Gangadharan N. A case for participatory disease surveillance of the Covid-19 pandemic in India. *JMIR Public Health and Surveillance*. 2020;6(2):e18795.
(46) Thomas G. Death in the time of coronavirus. *Indian Journal of Medical Ethics*. 2020 Apr 1(2):1–3.
(47) Biana HT, Joaquin JJB, The ethics of scare: COVID-19 and the Philippines' fear appeals. *Public Health*. 2020 Jun;183:2.
(48) Kapata N, Ihekweazu C, Ntoumi F, Raji T, Chanda-Kapata P, Mwaba P, et al. Is Africa prepared for tackling the COVID-19 (SARS-CoV-2) epidemic. Lessons from past outbreaks, ongoing pan-African public health efforts, and implications for the future. *International Journal of Infectious Diseases*. 2020;93:233–236.
(49) Kelly SE, Moher D, Clifford TJ. Defining rapid reviews: A modified Delphi consensus approach. *International Journal of Technology Assessment in Health Care*. 2016;32(4):265–275.
(50) Erwin C, Aultman J, Harter T, Illes J, Kogan RC. Rural and remote communities: Unique ethical issues in the COVID-19 pandemic. *American Journal of Bioethics*. 2020 Jul 2;20(7):117–20.
(51) Kelley M, Ferrand RA, Muraya K, Chigudu S, Molyneux S, Pai M, Barasa E. An appeal for practical social justice in the COVID-19 global response in low-income and middle-income countries. *The Lancet Global Health*. 2020 Jul 1;8(7):e888-9.
(52) Di Pietro L, Piaggio D, Oronti I, Maccaro A, Houessouvo RC, Medenou D, De Maria C, Pecchia L, Ahluwalia A. A framework for assessing healthcare facilities in

low-resource settings: Field studies in Benin and Uganda. *Journal of Medical and Biological Engineering*. 2020 Aug;40:526–34.

(53) Pecchia L, Piaggio D, Maccaro A, Formisano C, Iadanza E. The inadequacy of regulatory frameworks in time of crisis and in low-resource settings: Personal protective equipment and COVID-19. *Health and Technology*. 2020 Nov;10(6):1375–83.

(54) Piaggio D, Medenou D, Houessouvo RC, Pecchia L. Donation of medical devices in low-income countries: Preliminary results from field studies. In *International Conference on Medical and Biological Engineering*: Springer;2019, pp. 423–427.

(55) Stoeklé H.-C , Hervé C. COVID-19: Act first, think later. *The American Journal of Bioethics: AJOB*. 2020 Jul 2;20(7):W1–W1.

(56) Misra S, Le PD, Goldmann E, Yang LH. Psychological impact of anti-Asian stigma due to the COVID-19 pandemic: A call for research, practice, and policy responses. *Psychological Trauma: Theory, Research, Practice, and Policy*. 2020 Jul;12(5):461.

(57) Tessler H, Choi M, Kao G. The anxiety of being Asian American: Hate crimes and negative biases during the COVID-19 pandemic. *American Journal of Criminal Justice*. 2020;45(4):636–646.

(58) Mulu A, Bekele A, Abdissa A, Balcha TT, Habtamu M, Mihret A, Alemayehu DH, Beyene GT, Bayih AG. The challenges of COVID-19 testing in Africa: The Ethiopian experience. *The Pan African Medical Journal*. 2021 Jan 5;38(1).

(59) Maccaro A, Piaggio D, Dodaro CA, Pecchia L. Biomedical engineering and ethics: Reflections on medical devices and PPE during the first wave of COVID-19. *BMC Medical Ethics*. 2021 Dec;22:1–7.

Chapter 2

Cases Studies in Low-Resource Settings

COVID-19 Preparedness and Social Dynamics in a Sub-Saharan Africa Country, Benin[1]

Introduction

The SARS-CoV-2 disease (COVID-19) was declared a pandemic by the World Health Organization (WHO) on 11 March 2020, after a first wave of infections was reported in China at the end of 2019. Since then, it has become a global emergency and has continued spreading across the world. In Benin, the first case was declared on 16 March 2020 (1)). As of 25November 2020, Africa counted a total of 2.1 million confirmed cases and almost 51 thousand deaths, considerably less than other continents. Europe, in fact, totaled over 16 million cases and 370 thousand deaths, and North America over 15 million cases and 398 thousand deaths (2). There have been different attempts to explain this uneven distribution and morbidity of COVID-19, globally, including: the alleged genetic immunity of some ethnicities (2); the theory sustaining that some climatic conditions are inhospitable for SARS-CoV-2 (3)); the relatively young age of the population (2–4). Other theories claimed that the previous epidemics (e.g., Ebola) had strengthened low- and middle-income countries (LMICs) (5), making them more prepared for such outbreaks (2, 6, 7). Finally, there are the conspiracy theorists and deniers arguing that the pandemic is only affecting some world regions due to specific political reasons (8). Indeed, it is necessary to remember that the screening activities are not always comparable among countries (2). Thus, it is likely that the actual number of cases and deaths detected in the Global South is far from the numbers reported (9). As of 24 May 2021, Benin has reported 31 active cases for a total of 8,025 confirmed cases including 7,893 cured and 101 deaths.[2] These numbers have recently increased also due to the novel variants. In fact, as of 2 January 2022, Benin has reported 640 active cases for a total of 26,309 confirmed cases including 25,506 cured and 163 deaths. Nonetheless, these numbers still seem to show underreporting when compared to higher-resource settings (e.g., Italy has, as of April 2022, reported 1,253,056 active cases for a total of 15,106,066 confirmed cases including 13,692,608 cured and 160,402 deaths).

Since 2020, the pandemic has been affecting healthcare systems around the world, resulting in low-resource settings (LRSs) becoming the general norm, i.e., environments lacking means, specific knowledge, specialized personnel, medical

DOI: 10.1201/9781003436348-6

devices, and drugs, and with inappropriate medical locations. This has important ethical implications (e.g., the definition of scarce resource allocation criteria). In fact, while this situation was already familiar to LMICs, COVID-19 has overwhelmingly created LRS conditions in high-income regions, such as Europe (10). Nonetheless, significant differences among countries remain and even become more acute with regard to the social consequences of the pandemic. To this regard, in May 2021, Johri et al. (11) discussed the social consequences of COVID-19 on poor and vulnerable groups in India, arguing that in times of pandemic, the infection, prevention, and control measures (e.g., confinement) should be equipoised in respect to their impact on local societies, minorities, and economies. For this reason, and given the extraordinary social impact of the current pandemic, it was considered urgent to investigate the causes, starting from a survey on the social perception of COVID-19.

This article presents the results of this project, conducted by a multidisciplinary team involving sociologists, ethicists, and biomedical engineers from the Universities of Warwick, United Kingdom (UK) and Abomey-Calavi, Benin, in order to evaluate the population's perception in respect to COVID-19, the governmental measures and preparedness, and also concerning the Beninese healthcare system. The results of the interviews are grouped in main thematic areas (i.e., disease management, governmental measures, isolation, perception, pharmacological treatment, screening, and traditional medicine) that analytically show different aspects of the impact of the pandemic and its reception. Section 3 critically discusses these themes and the related surfacing emotion of fear with its dramatic consequences.

Methodology

The inductive method, leveraging several field observations to generate theories (12), was applied throughout this project. Figure 2.1 shows the phases of the project along

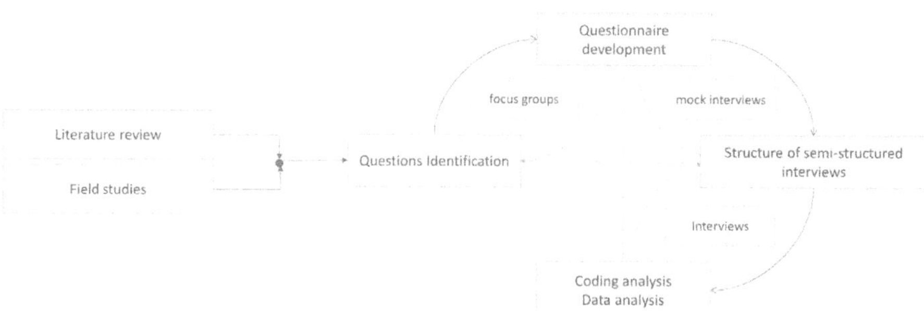

FIGURE 2.1
The various phases, applied methods, main objectives and outcomes of the project.

with their main objectives, applied methods, and outcomes. In particular, purposive sampling (13, 14), i.e., a non-random technique consisting of the deliberate choice of participants by the researcher according to the information they can provide based on their knowledge or experience (15), was selected and applied for any method requiring sampling.

This project largely relied on field observations, situational and field notes analysis, as well as semi-structured interviews to capture the opinions and perception on preparedness of part of the Beninese population in contact with the public and local communities. The cross-analysis *a posteriori* of such collected data allowed the generation of theories that could be compared with the relevant literature.

Literature Review

Relevant scientific literature was retrieved from databases such as Scopus using related keywords. Further literature was found applying a snowballing technique to the retrieved articles (16). Gray literature (e.g., the governmental websites, COVID-19 data, etc.) was retrieved by searching on Google.

Field Notes

Field notes taken during four field studies were used to support the analysis of the engineering aspects and of the preparedness of the healthcare locations, along with the related publications (17–19). Three of the afore-mentioned field studies took place in Benin (April 2017, January 2018, and November 2019) and one in Uganda (October 2019). During these field studies, the assessment of the conditions of medical devices and medical locations took place, as well as the ethnographic reconstruction of the local Beninese culture. Furthermore, issues related to the negotiation between local cultural particularism and the universalism of regulations and human rights, which are included in intercultural bioethics, have already been investigated and reported.

Semi-Structured Interviews

The first draft of the two semi-structured interviews, i.e., sociological and engineering, were created after a focus group with experts of ethics, bioethics, biomedical and clinical engineering, regulations and standards, and pandemic preparedness. The validity of the two surveys was investigated in two stages through two pilot-tests. Firstly, they were reviewed among the participants of the focus group for disambiguation and correction of content and language. Later on, it was tested with online semi-structured interviews, recruiting Beninese scholars not included in the former focus group. This was used as a further refinement of the semi-structured interview. The final questionnaire was written in two languages, i.e., English and French, as Benin is francophone and was held in French.

Interviewees were identified among workers at the Beninese Ministry of Health, biomedical engineers and technicians, healthcare workers, the police, anthropologists, public health experts, and civilians.

The full texts of the semi-structured interviews (available upon reasonable request by contacting the authors) were coded using NVivo 12-pro by two of the authors and were reviewed by a third author. The recurrence of the identified nodes (i.e., themes) along with their interrelations were analyzed to investigate and retrieve meaningful data.

The study received full ethics approval by the University of Warwick Biomedical and Scientific Research Ethics Committee (BSREC 140/19-20). Informed consent was sought before the start of the interviews.

On "Intersecting Cultural Domains"

Precisely, the goal of our study was not to fall back into anything resembling those debunked patterns of casting local, indigenous health cultural domains as "primitive" and "obstacles to Western modern practices". We consider that what was once called "*emic* perspectives" matters profoundly[3]. Quiroz and van Andel (20) underlined the social role of the utilization of plants in Benin, which cannot be simplified to the imperial dichotomies of "Euromodern vs. primitive" typical of the 20th century. Methodologically, this requires the research to "learn to see otherwise" (21). Similarly, Carolyn Sargent, in her ground-breaking medical anthropological studies of Benin, begs the question how "collaborative relationships" are possible in practical terms (22). Sargent allows us to see an even more interesting tension (23) than the merely epistemological and systemic conflict between Western (allopathic) medicine and any non-Western medicine.

These and related tensions are, in the case of newly emerging diseases, often expressed by indigenous populations in the Global as well as non-Western migrant populations in terms of a cultural ascription, as Decoteau (24) shows in the case of autism, as "Western" or "Global Northern disease". Accepting that it is not a single cultural factor that can lead to conflict and, thus, to problematic outcomes, such as we describe in the case of COVID-19, we take inspiration from Béhague et al. (25), that a proper understanding of the complexities at work and accepting cultural differences can also lead to *transformative agency.*

In other words, instead of merely looking at a problem presented by such complexity and finding a technical solution to solve or overcome the problem by negating a cultural factor, find *generative opportunities* within the complexities of the problem itself (as also described as a methodological paradigm in (26, 27). Along those lines, we find further inspiration in social research approaches that are inclusive of "seeing otherwise": drawing also on the Bourdieu (25), as well as the biopolitical critique of Michel Foucault, and the methodological purview of Critical Realism, Claire Decoteau has illustrated in her seminal research on HIV, masculinity, indigeneity, and normalizing, neoliberal biomedical rationalities in South Africa, how tensions between *emic* and *etic* perspectives lead to conflicts.

Results

The results presented in the following sections are a critical analysis of the semi-structured interviews, the retrieved literature, and our field notes. Unfortunately, due to COVID-19 restrictions, it was not possible to hold all the scheduled interviews. Consequently, a total of 13 interviews were held. Among the interviewees, four (30.8%) were healthcare workers, three (23.1%) public health experts, one (7.7%) hotel manager, one (7.7%) teacher, one (7.7%) hygienist, one (7.7%) anthropologist, and one (7.7%) policeman.

Nonetheless, the results are corroborated by a thorough analysis also referring to the relevant literature and field notes. Sections 3.1 to 3.3 present the results of our analysis stemming, in particular, from the literature review and the coding analysis of the interviews, which highlight the main surfacing themes and socio-cultural perspectives. In particular, the emerging confusion and skepticism surrounding COVID-19 will surface throughout the text. Similar perspectives were noticed in a successive work by the same authors (28) that aimed to investigate the social acceptability of COVID-19 vaccinations and tracking.

Literature Review: The Main Governmental Provisions

Online Resource 1 reports the main governmental provisions in Benin.

Engineering semi-structured interviews and field notes in

Benin, as in the rest of Africa, and most LRSs, has been and is facing different challenges, including:

- The lack of COVID-19 test kits, with the consequent under-detection of cases (29)
- The non-ideal screening of travelers and the different quality of screening dependent on the point of entry (29)
- The shortage of personal protective equipment and of healthcare workers and infrastructure (1, 30–32)
- The fear of stigmatization preventing ill people from going to the hospital (30, 32)
- Lack of access to adequate healthcare, housing, and sanitation (31, 32)
- Overburdened healthcare settings (30)
- Lack of adequate centers equipped for the treatment of COVID-19 (1, 29, 30, 32)

In particular, during our field studies in Benin we could assess the conditions of several hospitals, both private and public (18, 19). Indeed, a lack of essential medical devices and the relative spare parts, good health technology management system, preventive maintenance and of the tools needed for it (e.g., electrical safety analyzers and other analyzers), specialized doctors and biomedical engineers/technicians, as well as the harsh environmental conditions and unstable power supply, contributed to creating a very challenging environment even before the COVID-19 pandemic started (19). The situation was exacerbated with the start of the pandemic, as a more specific set of medical devices (e.g., ventilators), analyzers (e.g., gas flow analyzers), and expertise were needed. For detailed information on the presence of essential medical devices and their status (i.e., functioning or broken), please refer to Di Pietro et al. (19).

From the interviews, it emerged that although there is provision for some medical devices and equipment, the new logistics commission that oversees this does not communicate with the existing actors nor provides instructions or training.

One interviewee insisted on the cruciality of the training of paramedics, healthcare workers, and biomedical engineers/technicians about the new medical devices and technologies in order to

> "[...] ensure proper use and sustainability of the investment"
>
> *Health Inspector, female, 51 years old*

Finally, the interviewee stated that further technical actions are required, suggesting that putting up tents for the triaging and management of COVID-19 patients could be of great help.

Sociological Semi-Structured Interviews

Twelve people, two (16.67%) females and ten (83.33%) males, were interviewed. Their average age was 44, and their professional/expertise sectors are reported above.

The coding analysis resulted in the individuation of seven macro-themes, further divided into 45 themes. Online Resource 2 present the themes and their hierarchies. The discussed topics were: Perception of the disease (42%), Disease management (25%), Governmental measures (20%), Traditional medicine (7%), Screening (4%), Pharmacological treatment (2%), and Isolation (1%).

Governmental Measures

Since the WHO declaration of a state of pandemic, the Beninese government has distinguished itself for the implementation of an appreciable preventive strategy. In fact, it promptly started disseminating the first guidelines for the population, mobilizing international funds and human resources, requisitioning hotels to allow travelers to self-isolate, and strengthening the health and police system.

Indeed, Benin, like many African countries, welcomed the technical and financial support of the WHO, which allowed it to prepare adequately for the pandemic.

> "This translates into technical support (from the point of view of standards of care, installation of prevention devices, etc.) and financial. The WHO advises countries and mobilizes the resources of financial partners (World Bank, Canadian, Japanese funds, etc.), coordinates their interventions (weekly meetings

> of all partners), advocates, etc. without a mechanism of coercive power over the countries. The role of the WHO is much more in support than in decision-making [...]"
>
> *Epidemiologist, Male, 58 years old*

From a management point of view, the decision of establishing a sanitary cordon among the affected areas was promptly taken with the aim of reducing travel among the countries and regions most at risk. Furthermore, in order to avoid gatherings, it was soon decided to close the places of worship and to reduce the number of passengers for buses and taxis, and to impose the use of masks, even outdoors.

From a clinical point of view, the government soon aligned itself with the WHO indications regarding the preventive strategy and hygiene measures, which resulted in a non-negligible overall improvement in the equipment and hygiene conditions of hospitals and workplaces. Furthermore, the Beninese government soon decided to follow international guidelines regarding the tests for the screening of symptoms and the treatment. The former (i.e., PCR-TDR) are only performed in big cities, which can be very far from rural areas, apart from small teams that carry out home screening, or by telephone through a call center set up for this purpose. As for the treatment, it is mainly based on chloroquine, azithromycin, and vitamin C to strengthen the immune system. The communication of positivity to COVID-19 takes place after 72 to 96 hours on the phone or in person by an interdisciplinary team, which sometimes also includes a psychologist.

The complexity of this disease, being surrounded by a veil of uncertainties, highlighted the essential role of psychologists. In fact, during this pandemic, psychologists in Benin were recognized as the vital workers who were in charge of managing the communication of the positive outcome of the test, guaranteeing privacy, and managing fear.

Disease Management

The drug treatment is preemptively administered to those who have symptoms and are waiting for test results. In case of confirmation of positivity, it is continued for ten days, and on the 11th day the first nasopharyngeal swab is performed followed by a second one on the 13th day. If both tests turn out to be negative, the patient is considered healed and sent home with a certificate of healing.

The difficulty of screening lies in the fact that there are no symptoms directly attributable to COVID-19, apart from anosmia and ageusia, since fever or other flu symptoms may be attributable to other endemic diseases in Benin, such as malaria. Therefore, it is important to carry out specific tests, even though it is not always easy to reach the testing centers, wait for the slow turnaround of a non-computerized system, and deal with the scarcity of tests, drugs, and personnel in charge of the service.

> "Given the high rate of cases, the materials are not sufficient. In the specialized centers, there is an influx. The materials and beds are not enough."
>
> *Hygienist, Male, 36 years old*

As for drugs, moreover, an interviewee points out how initially Chinese-imported chloroquine at 500 mg was administered in hospitals. However, once these stocks were finished, chloroquine was divided into 250 mg doses. In fact, with the onset of COVID-19, Benin not only had the merit of tackling the scarcity of drugs and starting a local production of personal protective equipment, but also established low fixed prices to allow everyone to be able to purchase them without having to resort to the non-legal drug market sector, which is quite developed in Benin (33, 34).

> "[...] Still, the state has ensured that prices are kept at the same level in all pharmacies; the mask now I know it's at 100 CFA,[4] the chloroquine 500 mg was at 1000 CFA per pack"
>
> *Doctor, Male, 54 years old*

The scarcity of resources, as mentioned above, led to the occurrence of a situation universally shared by all countries which, in Benin as elsewhere, raised the bioethical question of the allocation of resources, locally dealt with prioritizing positive patients with comorbidities. Certainly, the economic, technological, instrumental, and organizational limits of the African countries, including Benin, represented an important bias that hindered and slowed down the response to the pandemic catastrophe, when compared to that of higher-resource settings.

Regarding this, some interviewees pointed out that the application of the WHO directives is not uniform in all countries. This recalls a very significant bioethical question, namely that of the generalist normative universalism, which claims to apply norms and precepts in all the contexts uniformly, without considering the specific situations in which they must be applied (10).

Traditional Medicine/ Endogenous Medicine

Benin is home to a traditional phytotherapeutic and herbal medicine[5] with such an ancient history and such a decisive value (34, 35) that it is also recognized by the Beninese Ministry of Health, which has established a National Program of Pharmacopeia and Traditional Medicine, which it relies on for the treatment of various diseases through preparations that are more readily available and affordable (36, 37). Also, for the treatment and prevention of COVID-19, the remedies of traditional medicine were soon resorted to (i.e., API COVID-19 and EB+ COVID-19). However, this alternative treatment did not gain much interest from the healthcare authorities (38).

These remedies were different from the traditional rituals that are linked to local animism (such as the consultation of the *Fâ*[6] and the consequent expiatory remedies) and which were also used in some cases. Rather, these remedies were based on the use of the same natural principles on which modern medicine partially relies, and could be a potential alternative in a context of scarce resources. Certainly, the inherent risk in the use of traditional preparations is that of the difficulty in controlling dosages, which could lead to dramatic consequences such as intoxication or death. Public health institutions would, therefore, benefit from scientific studies investigating the efficacy and safety of such traditional treatments. In fact, this use of local medicine could be an example of that necessary negotiation between normative universalism and situational contextualism, to be pursued with greater incisiveness.

> "Regarding this, there are a lot of herbal recipes that have a certain content of chloroquine within them. Traditional medicine should not be stigmatized, rather the active ingredients, their model of action on the body should be scientifically explored"
>
> *Doctor, Male, 62 years old*

> "Many use herbal medicines especially anything that is bitter; [...] artemisia prevents coronavirus [...]; others have used ginger, honey, lemon and turmeric, there is what they also call Kodo, they prepare it and drink it. [...] [others] believe that hot teas also prevent the coronavirus."
>
> "Consulting the Fâ and expiatory rites to ward off bad luck. In the middle of the Abomey plateau in southern Benin, the epidemic news with skin rashes and chickenpox are giving rise to expiatory practices".
>
> *Doctor, Male, 54 years old*

Perception of the Disease

Regarding the responsibilities of the government, the interviews conducted for this research reveal an excessive slowness in the communication of the test results, which evidently had consequences that cannot be underestimated, especially for COVID-19 positive patients and those who entered in contact with them, as well as a non-ideal management of people in quarantine.

With respect to the acceptance of measures related to the containment of the virus, many interviewees pointed out that travelers, who have been quarantined in hotels designated for this purpose, found themselves experiencing traumatic days due not only to the uncertainty about their health, but also to the neglect of the health service that was reserved for them. In fact, the results of the tests were delayed and the quarantine for many people was extended from 14 to 21 days. Moreover, many patients' (e.g., the elderly, pregnant women, etc.) needs were not taken into consideration. In addition, some interviewees highlighted financial perplexities regarding the destination of public funds that could have been used for health emergencies and not for the hotels to quarantine travelers. According to some, in fact, the decision to close the airports would have avoided these long-standing difficulties.

Many interviews revealed the malaise experienced by these travelers confined to hotel sites, ranging from feelings of frustration and discomfort to episodes of rebellion, sedated by the intervention of the police. Furthermore, it should be said, in general, that the interviews revealed a significant difficulty in complying with containment measures for the local population, who are particularly accustomed to shared sociability and sharing of daily rituals with the collectivity of the group (39).

The difficulties of the imposed isolation and the sanitary cordons that reduced the supply of raw materials in the country, and the consequent increase in prices, caused a surge in stress, which led to episodes of revolt and to an increase in domestic violence (40). Furthermore, the forced abstention from work had the effect of increasing boredom and a significant increase in alcohol consumption. What is surprising is that, although Benin has put in place a remarkable preventive strategy to increase its preparedness to manage the pandemic, it has never accompanied its security measures with relative legislative measures that guaranteed against any evasion by punishing offenders. In addition, Benin has not provided substantial support for workers who have been drastically affected by the isolation measures imposed by the pandemic. In particular, it is worth taking into consideration the informal economy sector which, in Benin, represents 97% of the gross domestic product and which was certainly among the sectors most affected by the economic and social consequences of the pandemic. This was underlined with insistence in the interviews along with proposals for direct support measures (through the distribution of food to the poor) and the reduction of taxes. The government has, up to now, intervened by offering economic support, in a single payment, to specific categories of workers considered most affected by the restrictive measures, which did not include workers in the informal sector (41).

Also particularly noteworthy is the sudden withdrawal by the state of all security measures, which gave way to a reliance on individual responsibility. This sudden step backwards, due to the balance between health imperatives and economic logic, received much criticism from the interviewees who are particularly confused by this behavior. In fact, at the very peak of the pandemic, the state authorized local elections, allowed children to return to school and the faithful to access places of worship, and removed the sanitary cordon, keeping in place only the measures of handwashing and use of a mask.

The weakening of the restrictions by the government had an impact on individual attitudes to the extent that individuals were led to believe that the danger announced by the WHO had been overcome.

This sudden change in the management strategy of the pandemic, in unison with the scientific uncertainties about the virus itself and the different interpretations of experts and scientists, was included in the categories of "infodemics" and "disinfodemics" by the WHO and UNESCO (42–45), which, with this newly coined lemma, also defined the consequent psychological condition of confusion and uncertainty generated on the population, as confirmed by most of the interviewees.

The initial fear of the population resulted in spontaneous comparisons with other communicable diseases spread locally, such as Ebola or Lassa fever, and the finding of a lower mortality and symptomatology from COVID-19 helped corroborate the feeling of underestimation of the pandemic event. In this regard, from the interviews it was possible to detect a particular caution shown by the Beninese authorities in the transmission of information, in particular in the dissemination of dramatic images that could have greatly affected people's sensitivity. This opens up some interesting hermeneutical questions: the reasons behind this choice, whether it was for protection or to avoid unpredictable reactions by the frightened population, both of which would be related to a questionable state paternalism, are unknown.

> "There are ways to communicate and impact. There are possible steps to follow in communication, this being a very important element in the management of the pandemic crisis. When Westerners showed pictures of infected people undergoing treatment, there was general panic. But here, we don't show any of that. There is nothing sensitive that the people are not afraid of. It is in communication that we must awaken people's awareness of respecting barrier gestures. And that's what I'm suggesting".
>
> *Anthropologist, Male, 39 years old*

As Abel et al. (46) argued, (dis)information overflow and political propaganda can make the situation more uncertain in respect to what public health measures should be accepted and followed. For future pandemics, there should be a more tailored approach according to which individuals should not be treated as objects. The compliance of individuals with public health measures depends on their proper communication and on the trust placed on the messenger. It is then clear how critical health literacy, giving an individual the ability to discriminate information, can play a crucial role in future pandemics and crises.

Sociocultural Representations

Epidemic outbreaks reveal logics that mobilize values linked to socio-cultural representations (47, 48). The latter are likely to tip the field of social interpretations of epidemic risk into a regime of doubt but also of vulnerability in connection with biomedical discourse (47, 49). These views are characteristic of the epistemological alignment of medical anthropology and refer to a dimension that goes beyond the limits of biomedical manifestations *stricto sensu* of epidemic outbreaks (50). For this reason, as for the perception of the current pandemic, different orientations can be schematically identified, which have certainly been corroborated by the uncertainty of the state attitude:

Adherence to the denier thesis

Adherence to interpretations related to traditional culture

Each of the two orientations has specific variants.

The denialist orientation is based on the scarcity of epidemiological data relating to the contagion in the country and reports the aforementioned hypotheses of the alleged African immunity (genetic, personal, and environmental) with respect to a disease that is attributed only to some populations, in particular to the Caucasian population. Opinions that the diagnosis or death from Coronavirus is unreliable and to be attributed to other diseases also contributes to denialism. This attitude of disbelief converges in the trivialization or underestimation of the pandemic that reaches its extreme with the adherence to full denial that, as it emerges from the interviews, seems to be very widespread in Benin.

> *"However, when you watch a white man approach the natives in the market, you find that they automatically pull themselves together. This perception, linked to the fact that the coronavirus is a white man's disease, causes users of the Dantokpa market to neglect each other [...]"*
>
> Anthropologist, Male, 39 years old

This denialist interpretative vein, together with the still very intense pain of the colonial domination, leads to the consolidation of some theses, already present with the advent of HIV, including that of the anti-birth conspiracy, according to which, COVID-19 is a lie used as a demographic weapon to control the numerosity of the population (51). This perspective matures in the background of the "clash of civilizations" which, as literature and history confirm, has fueled theoretical perspectives due to the "power of the imagination" and the "power of persuasion", typical of peoples who have suffered domination (52).

A more moderate version of this orientation is the one that sees in the current situation a form of neo-colonialism for which the West is planning a plot with economic ends (to increase earnings by selling drugs and masks) or an ideological "scam": Fearing dramatic scenarios of the future spread of COVID-19 in Africa, the West wishes to show its ready aid to countries in difficulty, thus perpetuating its dominion.

> "The very behavior of white people: we create things and then we come as a firefighter always for their own interests to the point that when there are serious problems, people trivialize: they say no, it's an invention. The government and their sides. It's not a real disease".
>
> *Epidemiologist, Male, 58 years old*

> "Which support the hypothesis of a 'conspiracy' around the imported masks".
>
> *Expert of Public health, Male, 60 years old*

> "We also note the denial: For many, the disease does not exist; it is a worldwide scam".
>
> *Policeman, Male, 33 years old*

The second line of interpretation stems from the traditional cultural background and does not only rely on the aforementioned therapeutic practices of traditional medicine, considering them more effective for African men than modern medicine, but also refers to animist spirituality. According to some, it would be possible to ascribe the etiology of COVID-19 to the contemporary dissolution of morality that threatens an apocalyptic outcome if there is no change in human behavior. Thus, consultations of the oracles were put in place, with the aim of researching the etiopathogenesis to calibrate the therapeutic-spiritual intervention based on it: alongside the herbal remedies, expiatory rites were celebrated, based on superimposition, typical of animist culture, between diseases and spells. In fact, according to this tradition, prevalent in the Fon cultural areas spread throughout the country, the disease is typically considered the empirical manifestation of the breakdown of the physical-metaphysical order due to moral reasons or "*envôutement*", i.e., jinxing (53). These beliefs lead us to consider the sick person as the bearer of the evil that is incarnated in them and that intends to strike the physical world through them. This implies that the patient is in most cases looked at with suspicion and considered susceptible to contagion, even if the pathology from which they suffer does not suggest any type of biological transmissibility.

Compared to today's COVID-19 situation, this stigmatization of the sick is clearly even more marked, since it is a communicable disease also from a biological point of view (54). The stigma therefore leads to keeping positive people at a distance, for the fear of a contagion not only physical, but also metaphysical. The interviews report a total marginalization of the sick, in particular the elderly and the most vulnerable (for example those who have other comorbidities) to whom the law has imposed isolation, because they are considered particularly at risk, but also of healthcare workers and of Westerners who are considered to be more exposed to contagion.

Desocialization is also applied to positive suspects or those who have recovered from the virus to the point that the government has provided for the issuance of "recovery certificates" for negativized patients who struggled to be readmitted into the social and work context. Stigmatization is one of the most dramatic social consequences of the pandemic, which brings with it the risk of avoiding being tested or self-reported for fear of social exclusion, and making the commitment of individuals and the government to contain the contagion useless.

On closer inspection in a context such as Benin, the economic and social consequences of the pandemic far outweigh the health consequences (consider the increase in illiteracy due to the closure of schools, the difficulty of accessing essential health services due to the overcrowding of hospitals).

For these reasons, it seems worthy to investigate the singular circumstances for which the dramatic extent of the consequences exceeds that of the spread of the virus. It seems that this is an ideological problem: it is as if the fear of the virus had caused more serious consequences than the virus itself.

The Emotion of Fear

Fear has always been a philosophical as well as a psychological word on which many contemporary theorists have developed important reflections: from heuristics, theorized by Hans Jonas, according to whom profiling a terrifying picture could activate in individuals the principle of responsibility (55) capable of promoting responsible actions to protect future generations, up to the reflections of the American philosopher Martha Nussbaum, a theorist of a social democracy based on skills, who modernly argues of the monarchy of fear (56), of escaping fear from the emotional dimension of the private sphere to enter the public scene, which also seems to occur during the so-called "COVID-19 era".

In fact, it seems that fear has led governments, such as the Beninese one, to spread an initial alarmism and to take drastic measures, consistent with those of the "first world" which, however, was experiencing a very different epidemiological situation. As a consequence, when the situation was compared with local data, it soon led to trivialization and denial with all the load of economic and social consequences mentioned above. Nussbaum, in this regard, observes that fear acts to cause alarmism and weakens capacity for judgment. Furthermore, as seen in the case of Benin, the fear of oppression, linked to historical resentment towards the West, has led to the identification of culprits, fueling undemocratic attitudes. Even in this respect, Nussbaum argues that the wrath-blame mechanism is typical of fear. Nonetheless, as seen in the reported case study, the fear of contagion has led to the exclusion and stigmatization of individuals and groups, setting up what Nussbaum calls policies of disgust. On closer inspection, however, fear is a primitive, narcissistic, and asocial emotion that is still causing more serious consequences in some countries than the spread of the virus itself. This arises when there is no adequate risk estimation.

Therefore, although it is extremely evident that science and its results must be universally usable, like health, which is a fundamental human right, it would seem appropriate to return to an effective risk assessment to better manage the consequences that derive from fear. This does not mean – as the Beninese government seems to have done – minimizing the current situation, particularly for already particularly vulnerable countries, but rather trying to close the knowledge gap through a greater diffusion of culture and scientific information that can allow a lucid and democratic analysis that activates responsible behaviors commensurate with specific situations, in compliance with that universality that is mindful of the particularistic and contextual differences that flow into the ethics of the situation. This means replacing fear with other emotions such as hope which, in Kantian terms, must be understood as a practical postulate and motor of action.

Conclusions

This study reveals technical, political, ethical, and sociological aspects of the preparedness to COVID-19 in a LRS, which, in their interconnection, show the importance of framing the issue with an interdisciplinary framework to offer a point of view that is complete and free from preconceptions. Furthermore, the study, although starting from the observation that the pandemic has put all countries on a universal level, namely that of the LRSs, leads to the conclusion that the hiatus among people and countries unfortunately still remains very strong: while high-resource settings are struggling to find strategies for the sanitary containment of the virus, the rest of the world is suffering the consequences of a virus that has barely appeared on its side. Therefore, it seemed appropriate to re-investigate the reason for the spread of such dramatic economic and social consequences in the face of a contained spread of the virus. It resulted that the ideology of individual and governmental fear has led to incorrect estimates of the risk that have been a harbinger of the aforementioned consequences. Ultimately, what emerges is that a generalized approach, even in the management of the COVID-19 pandemic, is always to be avoided because it leads to very serious consequences, such as those experienced in Benin. Rather than a one-size-fits-all approach, management that is tailored to specific contexts, situations, and circumstances, and that fosters diversity and inclusion, can provide a more effective and nuanced solution.

Notes

1 A. Maccaro, D. Piaggio, M. Vignigbé, A. Singl, L. Pecchia, *COVID-19 preparedness and social dynamics in a Sub-Saharan Africa country, Benin*, in «Health Promotion International», Volume 37, Issue 4, August 2022, daac105, https://doi.org/10.1093/heapro/daac105 (https://academic.oup.com/heapro/article/37/4/daac105/6674369?searchresult=1)
2 www.gouv.bj/coronavirus/
3 *emic* meaning here the internal elements and their internal consistency, resilience, and rationality as functional and functioning framework; *etic*, on the other hand, refers to an outsider's framework in the situation of cultural interpretation and interaction.
4 1 West African CFA franc = 0.0015 Euro
5 Although in this document the term "traditional medicine" will be used as it is the official name used by the government, "popular" or "endogenous" medicine, in the sense of "products drawn from its own cultural fund, experienced by society as an integral part of its heritage, as opposed to exogenous knowledge perceived as elements of another system of values" (Hountondji, P. J. (1994) *Savoirs endogènes: pistes pour une recherche*. Karthala, Paris, FR, ibid.), is a more suitable name.
6 *Fâ* is a divinatory art traditionally practiced by the populations of the Gulf of Benin.

References

(1) Osseni IA. Benin responds to covid-19: Sanitary cordon without generalized containment or lockdown? *Tropical Medicine and Health*. 2020;48:1–5.

(2) Musa HH, Musa TH, Musa IH, Musa IH, Ranciaro A, Campbell MC. Addressing Africa's pandemic puzzle: Perspectives on COVID-19 transmission and mortality in sub-Saharan Africa. *International Journal of Infectious Diseases*. 2020;102:483–488.
(3) Nguimkeu P, Tadadjeu S. Why is the number of COVID-19 cases lower than expected in Sub-Saharan Africa? A cross-sectional analysis of the role of demographic and geographic factors. *World Development*. 2020;138:105251.
(4) WHO. Africa media leader briefing on COVID-19. 2020.
(5) Keck F. Ebola, entre science et fiction. *Anthropologie & Santé. Revue Internationale Francophone D'anthropologie De La Santé*. 2015 Nov 25(11).
(6) Ayebare R, Waitt P, Okello S, Kayiira M, Ajok MA, Nakatudde I, Bhadelia N, Lamorde M. Leveraging investments in Ebola preparedness for COVID-19 in Sub-Saharan Africa. *AAS Open Research* 2020;3.
(7) Kapata N, Ihekweazu C, Ntoumi F, Raji T, Chanda-Kapata P, Mwaba P, Mukonka V, Bates M, Tembo J, Corman V. Is Africa prepared for tackling the COVID-19 (SARS-CoV-2) epidemic. Lessons from past outbreaks, ongoing pan-African public health efforts, and implications for the future. *International Journal of Infectious Diseases*. 2020;93:233–236.
(8) Mwaura W. Why some Kenyans still deny coronavirus exists. 2020. *BBC News*.
(9) Nakkazi E. Obstacles to COVID-19 control in east Africa. *The Lancet. Infectious Diseases*. 2020;20:660.
(10) Pecchia L, Piaggio D, Maccaro A, Formisano C, Iadanza E. The Inadequacy of regulatory frameworks in time of crisis and in low-resource settings: Personal protective equipment and COVID-19. *Health and Technology*. 2020 Nov;10(6):1375–83.
(11) Johri M, Agarwal S, Khullar A, Chandra D, Pratap VS, Seth A, Team GV. The first 100 days: How has COVID-19 affected poor and vulnerable groups in India? *Health Promotion International*. 2021;1:11.
(12) Goddard W, Melville S. *Research Methodology: An Introduction*: Juta and Company Ltd; 2004.
(13) Pope C, Mays N. Qualitative research: Reaching the parts other methods cannot reach: An introduction to qualitative methods in health and health services research. *BMJ*. 1995;311:42–45.
(14) Ritchie J, Lewis J, Nicholls CM, Ormston R. *Qualitative Research Practice: A Guide for Social Science Students and Researchers*: Sage; 2013.
(15) Etikan I, Musa SA, Alkassim RS. Comparison of convenience sampling and purposive sampling. *American Journal of Theoretical and Applied Statistics*. 2016;5:1–4.
(16) Marshall G. Snowballing technique. In *A dictionary of sociology*. 1998. pp. 201–216.
(17) Pecchia L. Assessment of medical devices in low income settings. In *Third WHO Global Forum on Medical Devices*, Geneva. 2017.
(18) Piaggio D, Medenou D, Houessouvo RC, Pecchia L. Donation of medical devices in low-income countries: Preliminary results from field studies. In Badnjevic A, Škrbić R, Gurbeta Pokvić L, editors. *International Conference on Medical and Biological Engineering*: Springer; 2019, May, pp. 423–427.
(19) Di Pietro L, Piaggio D, Oronti I, Maccaro A, Houessouvo RC, Medenou D, De Maria C, Pecchia L, Ahluwalia A. A framework for assessing healthcare facilities in low-resource settings: Field studies in Benin and Uganda. *Journal of Medical and Biological Engineering*. 2020 Aug;40:526–34.
(20) Quiroz D, van Andel T. The cultural importance of plants in Western African religions. *Economic Botany*. 2018;72:251–262.
(21) Decoteau CL. Learning to see otherwise. *Ethnography*. 2017;18:68–75.

(22) Chary A, Sargent C. Blending western biomedicine with local healing practices. *AMA Journal of Ethics*. 2016;18:691–697.
(23) Sargent C, Kennell JL. Elusive paths, fluid care. *African Medical Pluralism*. 2017;227.
(24) Decoteau CL. The "Western disease": Autism and Somali parents' embodied health movements. *Social Science & Medicine*. 2017;177:169–176.
(25) Behague DP, Kanhonou LG, Filippi V, Legonou S, Ronsmans C. Pierre Bourdieu and transformative agency: A study of how patients in Benin negotiate blame and accountability in the context of severe obstetric events. *Sociology of Health & Illness*. 2008;30:489–510.
(26) Stingl AI. The digital coloniality of power. Lexington, Forthcoming.
(27) Damien BG, Baxerres C, Apetoh E, Le Hesran J.-Y. Between traditional remedies and pharmaceutical drugs: Prevention and treatment of "Palu" in households in Benin, West Africa. *BMC Public Health*. 2020;20:1–14.
(28) Maccaro A, Piaggio D, Oronti IB, Vignigbé M, Gbokli A, Houngnihin R, Pecchia L. Social engagement in the fight against COVID-19 in the urban and peri-urban areas of Cotonou (Benin, Sub-saharan Africa): Acceptability of the vaccination and tracking program. *Frontiers in Medicine*. 2022 Jun 3;9:857890.
(29) Paintsil E. COVID-19 threatens health systems in sub-Saharan Africa: The eye of the crocodile. *The Journal of Clinical Investigation*. 2020 Jun 1;130(6):2741–4.
(30) Dubbink JH, Branco TM, Kamara KBB, Bangura JS, Wehrens E, Mac Falama A, Goorhuis A, Jørgensen PB, Sevalie SS, Hanscheid T. COVID-19 treatment in sub-Saharan Africa: If the best is not available, the available becomes the best. *Travel Medicine and Infectious Disease*. 2020 Sep;37:101878.
(31) Finn BM, Kobayashi LC. Structural inequality in the time of COVID-19: Urbanization, segregation, and pandemic control in sub-Saharan Africa. *Dialogues in Human Geography*. 2020;10:217–220.
(32) Post LA, Argaw ST, Jones C, Moss CB. Resnick D, Singh LN, Murphy RL, Achenbach CJ, White J, Issa TZ. A SARS-CoV-2 Surveillance system in sub-Saharan Africa: Modeling study for persistence and transmission to inform policy. *Journal of Medical Internet Research*. 2020;22:e24248.
(33) Ouendo EM, Makoutodé M, Paraiso MN, Wilmet-Dramaix M, Dujardin B. Itinéraire thérapeutique des malades indigents au Bénin (pauvreté et soins de santé). *Tropical Medicine & International Health*. 2005;10:179–186.
(34) Dresse A, De Baeremaeker D. *Amawoto. Le marché de la santé au pays du vaudou*: Agence belge de développement; 2013.
(35) Kpatchavi CA. *Savoirs, maladie et thérapie en Afrique de l'Ouest: pour une anthropologie du paludisme chez les Fon et Waci du Bénin*. 2011.
(36) WHO. *WHO Global Report on Traditional and Complementary Medicine 2019*: World Health Organization; 2019.
(37) Benoist J. *Médecine Traditionnelle et Médecine Moderne en République Populaire du Bénin*: Ecologie humaine; 1989.
(38) Hountondji PJ. *Savoirs Endogénes: Pistes Pour Une Recherche*: Karthala; 1994.
(39) Lévy-Bruhl L, Clare LA. *Primitive Mentality*: George Allen & Unwin Limited; 1966.
(40) Katana E, Amodan BO, Bulage L, Ario AR, Fodjo JNS, Colebunders R, Wanyenze RK. *Violence and Discrimination Among Ugandan Residents During the COVID-19 Lockdown* BMC public health. 2021 Dec;21:1–3.
(41) Stiftung FE. *Les Effets de la COVID-19 à Court, Moyen et Long Terme Sur Les Acteurs de l'informel*: 2020

(42) Posetti J, Bontcheva K. Disinfodemic: Deciphering COVID-19 disinformation. *Policy Brief.* 2020;1. Paris, France: UNESCO. https://en.unesco.org/sites/default/files, https://unesdoc.unesco.org/ark:/48223/pf0000374416.
(43) Younis M. COVID 19 and disinfodemic: A new age problem. *Pakistan Journal of Surgery and Medicine*. 2020;1:e248–e248.
(44) Hu Z, Yang Z, Li Q, Zhang A. The COVID-19 infodemic: Infodemiology study analyzing stigmatizing search terms. *Journal of Medical Internet Research.* 2020;22:e22639.
(45) Tangcharoensathien V, Calleja N, Nguyen T, Purnat T, D'Agostino M, Garcia-Saiso S, Landry M, Rashidian A, Hamilton C, AbdAllah A. Framework for managing the COVID-19 infodemic: Methods and results of an online, crowdsourced WHO technical consultation. *Journal of Medical Internet Research.* 2020;22:e19659.
(46) Abel T, McQueen D. Critical health literacy in pandemics: The special case of COVID-19. *Health Promotion International.* 2021 Oct 1;36(5):1473–81.
(47) Benoist J. Logiques de la stigmatisation, éthique de la destigmatisation. *L'information Psychiatrique*. 2007;83:649–654.
(48) Massé R, Weinstock D, Désy M, Moisan C. Perceptions populaires du risque et savoirs experts en contexte de pandémie: le cas du A (H1N1) au Québec. *Anthropologie & Santé. Revue Internationale Francophone D'anthropologie de la Santé*: 2011 Nov 29(3).
(49) Fribault M. Ebola en Guinée: violences historiques et régimes de doute. *Anthropologie & Santé. Revue Internationale Francophone D'anthropologie de la Santé*: 2015 Nov 25(11).
(50) Desclaux A, Sow K. «Humaniser» les soins dans l'épidémie d'Ebola? Les tensions dans la gestion du care et de la biosécurité dans le suivi des sujets contacts au Sénégal. *Anthropologie & Santé. Revue Internationale Francophone D'anthropologie de la Santé*: 2015 Nov 25(11).
(51) Alper S, Bayrak F, Yilmaz O. Psychological correlates of COVID-19 conspiracy beliefs and preventive measures: Evidence from Turkey. *Current Psychology*. 2021 Nov;40(11):5708–17.
(52) Schultz EA, Lavenda RH. *Cultural Anthropology: A Perspective on the Human Condition*: Oxford University Press; 2005.
(53) Dossi F. Le Bénin des grigris et des envoûtements. [The Benin of charms and bewitchments]. 2015. https://lanouvelletribune.info/2015/06/le-benin-des-grigris-et-des-envoutements/
(54) Egrot M. Renaître d'une mort sociale annoncée: recomposition du lien social des personnes vivant avec le VIH en Afrique de l'Ouest (Burkina Faso, Sénégal). *Cultures & Sociétés*. 2007(1):49–56.
(55) Jonas H. *The Principle of Responsibility. In* Search of Ethics for Technological Civilization: Libra; 2001.
(56) Nussbaum MC. *The Monarchy of Fear: A Philosopher Looks at Our Political Crisis*: Simon & Schuster; 2019.

Social Engagement in the Fight Against COVID-19 in the Urban and Peri-Urban Areas of Cotonou (Benin, Sub-Saharan Africa): Acceptability of the Vaccination and Tracking Program[1]

Introduction

The COVID-19 pandemic constitutes the most important public health challenge of our times. Globally, over the course of 2020 to 2022, health systems have been facing a great strain due to the lack of means and resources (e.g., priority medical devices [MDs], intensive care unit [ICU] beds, and COVID-19 tests and vaccines), as well as personnel and specialized knowledge. Consequently, even high income countries experienced a situation typical of low-resource settings (LRSs) (1–3). As of January 2022, the world experienced the third wave of the COVID-19 healthcare crisis, which was heavily influenced by the appearance of five variants of the SARS-CoV-2 virus (4, 5). The most effective solution identified was mass vaccination. For this reason, many companies promptly started working on vaccine development and testing. The genetic sequence of SARS-CoV-2 was published on 1 January 2020 and the first COVID-19 vaccine candidate entered human clinical testing with an unprecedented celerity on 16 March 2020.

To this end, the scientific community joined forces to develop the most promising vaccine candidates. Out of several vaccines developed on different technology platforms (6), only seven obtained the authorization for human administration.[2] Currently, more than 9 billion vaccine doses have been administered[3] worldwide. However, a great gap in coverage still exists between countries such that while some are already administering the third or fourth dose (e.g. more than 60% fully vaccinated persons in the United States), others are barely covered by the first dose.[4] Specifically, in high-income and upper-middle-income countries respectively, 71% and 73% of the population completed the vaccination cycle (as of 17 January 2022). Contrarily, in low- and lower-middle-income countries respectively, only 4.9% and

DOI: 10.1201/9781003436348-7

36% of the population received the vaccine. Overall, only 50.15% of the population completed the vaccination cycle[5] globally.

Unfortunately, inequalities and differences among countries are exacerbated by the cost of vaccines. However, cost is evidently not the only reason why vaccine deployment is unequitable among different countries. Other reasons include governmental preparedness, poor transport links, and inability to maintain a cold chain, which is vital for preserving some vaccines. This situation resulted in governments, scientific institutions, and healthcare professionals embarking on unprecedented collaborations to promote equitable access to COVID-19 vaccines through speedy deployment and distribution in low- and middle-income countries (LMICs). This materialized in the COVAX initiative, co-led by the Gavi Alliance, the Coalition for Innovations in Epidemic Preparedness (CEPI) and the WHO.[6] The vaccine distribution strategy references an ethical framework established by the WHO's "Strategic Advisory Group of Experts" on Immunization (SAGE) (7–9). Nevertheless, these efforts did not automatically translate into popular support for COVID-19 vaccines worldwide (10), and particularly in LMICs (11, 12).

One of the main reasons for hesitation in relation to vaccination is the lack of clear information from the government and scientific institutions, exacerbated by the circulation of inaccurate news through social media and the internet (13). This phenomenon, referred to by the WHO as "infodemic"[7] (14–16) or "disinfodemic" (17), has particular repercussions in LMICs, where it feeds widespread negationism (18). Despite the COVAX target to provide up to 600 million COVID-19 doses by the end of 2021, and the fact that 52 African countries received almost 177 million doses (19) of vaccines as of January 2022, the vaccination target for Africa was not reached. The authors of this paper believe that this may not be linked to a lack of supply, but rather, to the local culture and pessimistic approach relative to the vaccines.

This study was limited to Benin since it is inhabited by a melting pot of ethnic groups, is considered the "cradle" of the Vodun religion and an exemplar for the traditional culture dominant in Sub-Saharan Africa, and is also enrolled in the COVAX initiative that saw the deployment of a specific National Vaccination Plan.[8,9] A total of 26,036 confirmed COVID-19 cases with 162 deaths were reported to the WHO in Benin from 3 January 2020 to 13 January 2022, while no less than 1,897,214 vaccine doses had been administered as of 3 January 2022. This means that 11% of the population completed the vaccination cycle, and 15% received at least one dose.[10] Interestingly, according to available literature (20, 21), Benin is one of the countries with the lowest COVID-19 vaccine acceptance rate. In general, the reported data related to the spread of COVID-19 in Africa are strictly dependent on tracking systems, which are evidently problematic in LMICs due to institutional and technical under-preparedness (22, 23). Therefore, an in-depth analysis of the social reasons for these phenomena is essential to understanding the underlying problems in order to favor a better vaccine rollout and acceptance. It is important to note that the COVID-19 vaccine was introduced *pari passu* with the rich tradition of healthcare treatments that offered alternative herbal-based therapies (EB + anti-COVID-19, Api-COVID-19, etc.), which were well received by the population (24–26).

This work presents the results of a multidisciplinary collaboration which focused on the analysis and critical appraisal of the perceptions of the Beninese population and the novel challenges brought about by the COVID-19 pandemic. It relies on a previously created *hermeneutic heuristic framework* that combines the theoretical structure of intercultural bioethics and the empirical, inductive, and contextualized approach (by design) of the sciences (in this case biomedical engineering), thus proposing solutions inspired by the concept of "frugality" that takes account of the particularism of each context, while tending towards a model of universalism. Part of this framework relating to the frugal design of medical devices is already published (27), while the other part is currently under consideration. Our research goals included capturing popular perceptions about COVID-19 and strategies put in place for fighting the pandemic (i.e., test and trace, vaccines), investigating the main reasons behind the emerging local perspectives, highlighting the main local reactions, and understanding the differences between the perspectives and reactions of the inhabitants of urban and peri-urban areas. Given this background, the results of the study will be a proxy to understanding the overall level of commitment in the fight against COVID-19, and to support community engagement for the future of public health management in LMICs.

Materials and Methods

This article is the outcome of an international multidisciplinary collaboration between the ABSPIE Lab of the University of Warwick, United Kingdom, and the LAMA of the University of Abomey-Calavi, Benin, that brought together engineers, sociologists, anthropologists, and bioethicists with the common goal of bringing key interdisciplinary issues that affect LRSs to the attention of the world. The project was divided into three phases, mainly; (a) the hermeneutical (interpretative) phase – conducted as a field study by the researchers of the University of Abomey-Calavi, (b) the analytical phase, and (c) the synthetic phase. The last two phases were led by researchers at the University of Warwick.

In light of the aims of the project, a qualitative methodology and inductive approach were deemed appropriate and selected for the first phase. The research methodology was structured around data collection methods (semi-structured interviews and focus groups), with experts drawn from groups specified in the criteria contained in the government's vaccination strategy as dictated by the COVAX initiative, i.e., the healthcare sector, traditional local medicine practitioners (*tradithérapeutes*), as well as representatives of the local people, in particular, carriers of comorbidities, people aged 60 and above, and vaccinated subjects. The research team prepared the layout of two questionnaires (attached in the supplementary materials), which formed the basis for semi-structured interviews (held either in French or in Fon) that took place between June and August 2021,[11] while also taking into account the writing of the concept note, interviews, translation and transcription of data, and thematic sorting. The fieldwork was conducted by two researchers who assumed responsibility for data transcription and thematic sorting.

Survey populations were selected using the reasoned choice and snowball techniques. Reasoned choice provides that the sample is chosen in a way that represents the studied population as accurately as possible. For this reason, all the groups involved in the vaccination campaign were captured. The sample size was determined by the saturation threshold, i.e., the number of interviewees was stopped when no new codes or themes emerged from the interviews, rather, the same ones started recurring (28, 29). The final sample size was 34, including nine comorbidity carriers, six healthcare workers, eleven *tradithérapeutes*, and eight people over 60 years old. Particular attention was paid to the gender dimension in the conduct of the interviews to make sure that it was as balanced as possible. Overall, we believe that this mixed-background population could be a good representative sample of the Beninese population. Table 2.1 reports the characteristics of the interviewees.

Finally, the survey data were recorded and then transcribed for the second stage of thematic sorting. When necessary, the transcription was preceded by translation for interviews and passages conducted in local languages (especially French and the Fon dialect). In addition to the interviews conducted, direct observation was chosen as an additional technique for collecting data that could not be captured by speech. The research site was initially identified as Cotonou, but was later expanded to Abomey-Calavi and Sèmè-Pkodji – two dormitory cities adjoining Cotonou. This choice is justified by the fact that the latter two are affected in the same way as Cotonou in terms of the prevalence of COVID-19, in addition to being part of the "red zone" of the pandemic. These choices thus reflected the social urban perception of vaccination, more so because Cotonou and its peripheral towns (Abomey-Calavi and Sèmè-Pkodji) have 38 vaccination sites out of 78, totaling almost half for the whole country (76 cities).

In the second phase of the project, the results of the field study was analyzed by the University of Warwick team using thematic analysis, triangulation of data collection sources, and content analysis techniques, which was corroborated with a broad literature review. The last (synthetic) phase, in the Hegelian sense, involved combining the results of the first two phases, thus using the framework for proposing

TABLE 2.1

Description of the characteristics of the interviewees.

Average age	**48.76 (29 – 89 years old)**
Gender	61.8% Male; 38.2% Female
Category distribution	17.65% Caregivers
	29.41% with comorbidity
	20.59% over 60 years old
	32.35% *tradithérapeutes*

strategies and solutions for addressing the identified problems. The framework aims at bringing to public attention the problems identified and interpreted in a multi-perspective way, and this paper is one of the results and part of the ongoing dissemination strategy.

Results

The conducted interviews revealed many elements worthy of interest related to the social engagement of the Beninese population in the fight against COVID-19. The use of the hermeneutical framework made it possible to identify four macro-areas: a) governmental and technical aspects, b) social aspects, c) traditional care aspects, and d) ethical aspects. It was from these four areas that crucial elements emerged. Figure 2.2 represents a tree diagram summarizing the main themes pinpointed during the coding of the interviews.

Governmental and Technical Aspects

The interviewees were requested to give their opinion on the governmental management of the COVID-19 pandemic, including the vaccine delivery mechanism system. The respondents report that the vaccination system followed in Benin is in line with the protocols established by the WHO (30), which registered a good system of vaccine conservation and hygiene in the administration and disposal of waste. However, two

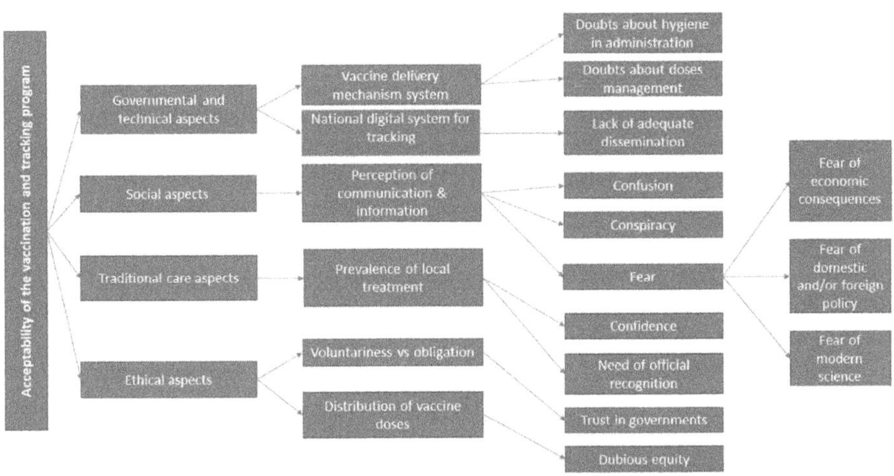

FIGURE 2.2
Tree diagram highlighting the main themes surfacing from the interviews.

accounts were opposed to an overall good opinion of the mechanisms put in place. An interviewee declared his concerns about the hygiene and procedure followed.

> "The biomedical devices in place do not always meet the technical needs of the vaccination sites. Here at the National University Hospital Centre, for example, it is not possible to observe the distance in case of turnout. Also, there is no observation room to allow those who feel ill after vaccination to rest (…). The second part I have doubts about is the vaccinator. I don't know if he or she follows the precautions. In principle, after a vaccination, the vaccinator must change gloves before vaccinating another person. But I'm not sure if he respects this standard because they don't have enough gloves. After four or five vaccinations, the vaccinator uses hydro-alcoholic gel. So, they get up from time to time to wash their hands."
>
> *Caregiver, male, 54 years old, Cotonou, 09/08/2021*

Another respondent expressed concerns that temperatures to keep vaccines viable may be subject to fluctuations.

> "So, the question is, do we have all the equipment we need to respond? That's the question that came up. I'm already telling you that we don't have a refrigerator in the infirmary and that we put the vaccines in the cooler in the morning until maybe 3 p.m. before vaccinating people, maybe that's a guess, that's my point of view. So, in terms of safety, maybe we're breaking the cold chain a bit."
>
> *Health worker, 44 years old, Cotonou, 23/07/2021*

Most respondents report that they are aware of the system of registration of vaccinations that passes through a national digital system that releases a QR code and is connected to a mobile app.[12] However, some people are unaware of this and report on some procedures that are still paper-based and should be followed by subsequent online data entries.

Ultimately, the Beninese government's handling of the pandemic is perceived in a controversial way. Those who appreciate technical health management are pleased with the government's agenda and attribute its merits to the WHO, while those who are more critical, instead, look favorably on the experiences of other states that have used traditional medicine in the fight against COVID-19 (31) (e.g., the Malagasy president). Moreover, the interviewees show many doubts about the Beninese local government's ability in overcoming the resistance to the vaccination drive due to accusations of corruption (e.g., collusion for economic reasons with the "whites", poor commitment to the recognition of local traditional medicine, and faking the vaccination procedure), lack of clarity, and disinterest in the real involvement of the population.

> Before proceeding to vaccination, the leaders must seek public opinion on the ins and outs of the vaccine and in relation to the rumors that are circulating about the

> vaccine. Today, Africans have a very high level of awareness and no longer want to submit to the yoke of white imperialism. (…). On the other hand, if it were with our plants and barks, I know that such and such a plant cures such and such a disease, and thus I will be able to submit to the indicated treatment.
>
> *Person over 60, female, 89 year old, Cotonou, 27/07/2021*

> Now our leaders are being vaccinated to set an example. What is being inoculated into their bodies? We are not doctors to know. What is to be administered to the rest of us as well, we have no idea. Sincerely, they have love for their citizens, let them not be baited by money to feed us.
>
> *Person over 60, female, 89 years old, Cotonou, 27/07/2021*

> The lack of information means that few people are currently vaccinated. Only intellectuals who have to travel and those who are among the targets listed by the government do so. I can say that the communication around vaccination is not going well. There is no mobilization. Apart from media releases, there is no other form of social mobilization.
>
> *Caregiver, male, 55 years old, Cotonou, 06/07/2021*

Social Aspects

The interviews sought to highlight the position of the population in relation to the COVID-19 pandemic and the reasons that justified them. There is clearly a misconception of the true nature of COVID-19, which gives rise to an overall rejection of the vaccine. Conspiracy theories are very common, and conspiracy against the population is foreseen and explained with the tradition of *zangbeto*, in the local culture the victim *par excellence*.[13]

> This story talks about the *zangbeto* that is caught in a mouse trap. In this story of the COVID-19 pandemic, there is something fishy going on; something is being plotted to harm 'Man'. Something is being hidden from us. We are not being told the truth.
>
> *Person over 60, female, 85 years old, Cotonou, 28/06/2021*

The emotion that is recorded in a more recurrent way is the *fear* that feeds the denial of COVID-19 and the rejection of the vaccine.

> Some people say that if you get the vaccine, you will die in two years. Maybe that's why people refuse to get vaccinated. Who wants to die? So, people are suspicious. This is the feeling that drives the population; they are afraid. People doubt who is telling the truth and who is not. And so, they decide not to be vaccinated and to remain as they are.
>
> *Comorbidity carrier, female, 58 years old, Abomey-Calavi, 23/07/2021*

The *confusing information* bombarding the population through communication networks and social media (about post-injection manifestations and the doubts of the government about the age requirements for the vaccination) is the main cause for fear. Social media builds up a real "mobilization strategy" made of WhatsApp messages that are spread by unknown numbers through message chains or forums on the internet. Because this information source is not controlled and/or censored by the authorities, they are considered to be confidential and, therefore, more reliable.

> At the moment, I am not willing to do the vaccination. I am willing to do it if it will protect my health. But, with the rumors on social networks and the alerts on TV channels, and with what is recorded here and there after injections – the illnesses that it causes afterwards, I am not tempted to do it immediately.
>
> *Comorbidity carrier, male, 32 years old, Cotonou, 22/06/2021*

The declared fear can be traced back to three major issues: fear of economic consequences, fear of politics, and fear of modern science.

Fear of economic consequences: This appears to be the major reason for denying the existence of the virus, especially in rural areas that are mostly affected by the restrictions (such as the ban on having markets).

> I have a brother in the village who has a cough. He suffers from a stubborn cough and he is a prominent gentleman in the community who enjoys a certain socio-economic prestige. Eventually, at the local health center, he was found to be a carrier of the COVID-19 virus and had to restrict himself to quarantine requirements. He saw all his activities interrupted, the health center that took care of him is closed to users, and the village itself is stigmatized with the adverse effects of poor sales for traders of doughnuts, and especially mustard and Akassa ball, which the locality is known for because of its quality of production.
>
> *Comorbidity carrier, female, 55 years old, Cotonou, 29/06/2021*

Fear of domestic and/or foreign policy: As anticipated, the population revealed that they have no trust in the government, based on claims of corruption, and are therefore fearful of any government action. In addition, the fear of politics is also relevant in the international context, and traces its roots to the colonial experience in the past centuries: It is believed that COVID-19 would allow foreign rule to return, thus controlling African populations, or even decimating them in order to limit any anticipated danger that could arise from the increasing population size.

> The anti-COVID-19 vaccine is being introduced in our countries in Africa to shorten the lifespan of the population. The great powers of this world consider that we are overpopulated, there are more people than we need in the world and therefore we have to find a way to reduce the world population size, especially in Africa, to a given threshold. I suppose that we must be six billion in the world and here we are at ten billion! The surplus is thus to be removed.
>
> *Person over 60, male, 67 years old, Cotonou, 19/07/2021*

Fear of modern science: The population adduces a number of reasons to justify the fear of modern science, coupled with politics. They feel this is another cornerstone of the above-mentioned "anti-African" plot. In particular, some interviewees underline the small number of known COVID-19 infections in Benin and the high temperatures in Africa, from which they deduce that it is a disease that does not particularly develop at such climates and does not particularly affect the Africans' genetics. In addition, with regard to the fight against COVID-19, the population is suspicious of the speed at which vaccines were trialed and believes that this could compromise its effectiveness and safety. In this regard, other respondents reinforce the hypothesis that it is a political-economic expedient because they are dubious about the reasons for the speed behind vaccine production and spread in spite of the fact that other more common diseases in Africa are still waiting for a cure (e.g., malaria or AIDS). This also brings up the suspicion that the African population is being used as "guinea pigs" to test certain types of vaccines. All these reasons reinforce the general rejection of not only the COVID-19 vaccine, but also other vaccines that were accepted in the past.

> COVID-19 is not in our country. It is the white people who brought it.
>
> *Tradithérapeute, male, 35 years old, Cotonou, 16/06/2021*

> The whites do not like blacks. With the number of deaths caused by COVID-19, they do not take action to save themselves and they want to save us. This way of acting should make black people think. AIDS has been around for years without a vaccine, and it's COVID-19 that they have found a vaccine for. I don't want this vaccine!
>
> *Tradithérapeute, male, 35 years old, Cotonou, 16/06/2021*

> In the face of every disease, Africans are used as test drivers. They use us as test subjects to improve their products. It is true that the whites have already started taking the vaccine. But the drugs they use are different from the ones they send to us in Africa. The quality of the pharmaceutical products is not the same. That's why our Heads of State go abroad for treatment when they are ill.
>
> *Tradithérapeute, female, 33 years old, Cotonou, 16/06/2021*

> As a result, I have had to dismiss the polio vaccinators who go to households door-to-door because we don't really know any more if they are polio vaccinators or if it is the COVID-19 vaccine.
>
> *Tradithérapeute, male, 35 years old, Cotonou, 16/06/2021*

These convictions nourishing the social perception of the ongoing pandemic are extremely acute in the peri-urban and rural areas where there is a widespread belief that the virus is fictitious. Also, the distance from the vaccination mechanism, which is concentrated in urban areas, and the government's disregard with respect to the control of compliance with safety measures contribute to these beliefs.

> The disease is not in our villages. The villagers ask us to be careful. They ask us not to come and contaminate them. Because the disease is not in the village. It's all fabrications when they tell us that there are cases of COVID-19 in the villages.
>
> *Tradithérapeute, male, 34 years old, Cotonou, 16/06/2021*

> In the villages, compliance with these rules was lower. But in our urban areas, the police will arrest you if you don't respect the barriers.
>
> *Tradithérapeute, male, 34 years old, Cotonou, 16/06/2021*

> It is in Cotonou that we can say that people are more or less aware of COVID-19 restrictions, but when they leave Cotonou, it is not so evident. (...) In Cotonou, you can't go into a pharmacy without a mask, but in Parakou, for example, you can go into a pharmacy without a mask, which reduces the motivation to get vaccinated.
>
> *Comorbidity carrier, male, 43 years old, Sèmè-Pkodji, 26/07/2021*

Traditional Care Aspects

In Benin, there exists a rich context of traditional medicine based on the use of plants for self-medication and the multiplicity of traditional healers (i.e., *tradithérapeutes*), many of whom are recognized by the healthcare system, while others are considered charlatans. The main problem of traditional medicine is that there are no scientific studies that establish doses and confirm full alternatives to modern treatments. Nevertheless, there is a great engagement of the population (including medical doctors) in this type of treatment. The interviewees reveal the strong belief that traditional medicine is a valuable preventive prophylaxis tool against COVID-19, and think it is better than any other modern treatment.

> We have strengths in traditional medicine that can help us prevent and cure even COVID. These are the recipes and I believe in them. We are told that COVID-19 is like the flu that chloroquine can cure. But we also have drugs that play the role of chloroquine and Azithromycin. So the problem is solved traditionally.
>
> *Caregiver, male, 54 years old, Cotonou, 09/08/2021*

> Do you see in the West, despite the development in their health systems, that their populations have suffered a hecatomb? Have you not asked yourself why not here? It is precisely because of our food and pharmacological habits based on plants! These people, we must not give in to ignorance and simply follow them. They will simply kill us. No, in Africa we have everything we need to cure all kinds of diseases.
>
> *Tradithérapeute, male, 66 years old, Abomey-Calavi, 23/07/2021*

The respondents are aware that studies of traditional medicines require time and investment of money. Respondents report that some local products are licensed, while others have been banned. This gives rise to the fear of government repercussions against the traditional healers. In fact, based on some reports, the government tried to involve traditional healers in the management of the pandemic by creating a group of experts in traditional medicine. This, however, did not give them the opportunity to take initiatives. In reality, local governments are acting very cautiously with respect to local medical traditions, fearing international repercussions (as in the case of Madagascar, where the president sponsored local treatment against COVID-19, but later withdrew the regulations following pressures from the international community).

> The problem of complementarity is primarily a political and geostrategic issue of the great powers. To date, no blacks have been allowed in the circle of researchers conducting research on COVID-19. Without protection, we cannot exhibit our products against COVID-19 because we will be fought. The case of the Malagasy President is an example. The second example is that of Dr. Agon Valentin. Burkina Faso had accepted his product, API-VIRIL, as a remedy against COVID-19. However, when the 'Whites' got involved in the dance, his product was banned from use against COVID-19, and was withdrawn from the Burkinabe market and banned from sale in his country, Benin.
>
> *Tradithérapeute, female, 37 years old, Cotonou, 16/06/2021*

> In fact, there are many problems at this level. First of all, there is the problem of the non-complementarity of the biomedical and endogenous offer. Moreover, the promising endogenous physicians have difficulty in getting together to produce quality medicines. The quest for individual recognition leads them to evolve in a scattered manner. As an example, to face the COVID-19 pandemic, the Beninese government has set up a committee of experts. This committee met with the traditional medicine actors that we are. It asked us to organize ourselves to make proposals for a remedy against the pandemic. Instead of uniting to produce something of quality, each healer went to the committee of experts with their product. It was so disorganized that the expert committee has not made a decision to this day yet. So, it's an internal problem for us.
>
> *Tradithérapeute, male, 34 years old, Cotonou, 16/06/2021*

Fear recurs as the dominant theme in the interviews, and one of the remedies to fear is considered to be traditional medicine. In fact, some respondents report that before vaccination against COVID-19, some people had undergone a spiritual purification and made specific sacrifices to local deities, or consulted the oracle of local geomancy. This represents another side of traditional medicine, not necessarily related to herbal treatments.

> There are some who are vaccinated and have not had any effects. But when I approached them, they told me that they prepared themselves psychologically, medically, and spiritually by giving themselves to God to become immune to COVID-19. They too were afraid and took steps to overcome the fear. A few days

before taking their doses, they took chloroquine and azithromycin to fight against the virus and paracetamol to prevent fever, as is done for children during vaccination. They also prayed a lot and did not want their husbands and wives to follow suit. The others did it first to see what it would do.

Caregiver, female, 45 years old, Cotonou, 12/08/2021

There is also that deity '*Sapkata*[14]'. Let us not forget it. It is '*Sapkata*' who governs all diseases. If one makes the required sacrifices to her, she protects and prevents the population from falling ill.

Tradithérapeute, male, 34 years old, Cotonou, 16/06/2021

All illness comes from a spiritual imbalance that negatively impacts the physical condition. Therefore, if we can regulate all spiritual imbalances, no one will fall sick because illness is a spirit. It is enough to make *Soudjo* conjure the spirits responsible for the disease so that they do not come into our environment.

Tradithérapeute, male, 34 years old, Cotonou, 16/06/2021

Moreover, we had consulted the *Fâ* oracle about the origin of COVID-19. [...] The *Fâ* was clear and precise. He even said that young people will not be victims of this disease. Following these revelations, we made sacrifices. Only charlatans paid money to make the sacrifice against COVID-19. We should then do the ritual called *tokplopklo*. But due to lack of funds, this has not yet been done. The *tokplopklo*[15] is done by several people. The government itself has to give money. But it refuses to provide the necessary means. This is proof that our leaders do not love their country.

Tradithérapeute, male, 66 years old, Abomey-Calavi, 23/07/2021

In addition, we have other methods to send the disease away from our lives. These include taking *adjanouhlahoun* and *zozoman*. When you take a bath with the potion of these two leaves, any disease stays away for a good period of time.

Tradithérapeute, male, 34 years old, Cotonou, 16/06/2021

Ultimately, it emerges that the population feels more represented by traditional medicine than by modern medicine and would very much appreciate the formalization of the former. It is also evident that an official involvement of traditional medicine in the fight against the pandemic would lead to a stronger popular engagement.

But I think the authorities need to approach our traditional healers to learn about the recipes they use, test them scientifically before accepting or rejecting their use. If not, we will lose. Besides, we are used to losing. Because our old people are dying every day with their recipes. The best strategy would be to accept

> these traditional plants, study them scientifically and make them available to the population. Researchers are there for that. Our governments must change their positions. They know full well that traditional medicine is full of qualities. So we should not categorically refuse without experimenting.
>
> *Caregiver, male, 55 years old, Cotonou, 06/07/2021*

d) Ethical Aspects

In Benin, vaccination is voluntary. In the first phase, it was limited to some categories (i.e., health professionals, people with serious comorbidities, and those over 60 years old), but now, it is open to people aged 18 and older. A specific section of the interviews was dedicated to asking how the population considered the theme of mandatory health treatments and if the vaccine against COVID-19 was perceived as an obligation. We discovered that many healthcare workers were also opposed to vaccination, although they were not obliged to get it done at that time.

> So far, nothing! I thought that if someone doesn't get vaccinated, they won't be able to do their job, that they will be fired, but so far there is no such pressure. Vaccination is not mandatory for healthcare workers. But if it were compulsory, I would make one condition: I have a medical certificate describing my state of health and I would write to the authorities that it's against my will and that I'm getting vaccinated to save my job.
>
> *Caregiver, female, 45 years old, Cotonou, 12/08/2021*

Regarding the COVID-19 passport, the population is opposed to it and considers it as a form of oppression.

> It is wrong to talk about the COVID-19 vaccination passport. But we are not going to let it happen. However, it's a pity.
>
> *Tradithérapeute, male, 34 years old, Cotonou, 16/06/2021*

Finally, the interviewees underlined the lack of equitability in the distribution of vaccine doses: the sharp difference between areas of the world prevents a global community vision.

> What did not work was the international community. It does not have that kind of solidarity. If everyone is affected, everyone should benefit equally from the vaccine doses. But today we see that the Westerners give priority to themselves first, and then send us a few drops. That didn't work. The mobilization of resources has not worked. I'm sure that the resources that countries have received and mobilized have not allowed them to do much.
>
> *Comorbidity carrier, male, 43 years old, Cotonou*

Discussion

The fight against COVID-19 will be successful only if it will build on a complete engagement of the population. As shown in some of the reported interviews, the latter is strictly connected to many sociocultural aspects. The first aspect is the net gap between the population and politics, which often relies on contradictory and unclear communication strategies for the population, thereby fomenting conspiracy theories and negationism. Overall, it seems that local policy is geared more towards complying with international standards rather than towards assessing and tackling what is happening in the country.

Firstly, Benin is carefully trying to follow the WHO protocols of conservation, administration, and disposal of waste, but does not deal with the very serious risks that could result from blackouts or unstable electrical supply to the refrigerators containing vaccines. The cold chain (i.e., cold storage during the delivery and storage at the provider's facility up until the administration to the patient) is required by some vaccines to maintain their potency and effectiveness. This may be very difficult to maintain in a LRS such as Benin, which has an electrification rate of 43% and a quality of electricity rated as 2.06/7 (as a comparison, Italy has a 100% electrification rate, and a quality of 5.91/7). The quality of the electricity is compromised by frequent blackouts, sags and swells, and the frequent unavailability/breakdown of generators, as also reported in previous studies (32, 33). The more general concern is that for politics, economics, and in the present case, also health, there are established international standards, which although have the benefit of standardizing practices and facilitating trade between countries, are in fact limited by not considering the variety of specific contexts and situations. The alleged universality of standards is often generic and risky because, as in the case of the storage temperature of vaccines, it could call into question the effectiveness and safety of health technologies and, consequently, the health and rights of patients.

The second evidence of the policy's distance from the population is that there is lower control over distancing measures and spread of vaccination and its tracking mechanism, particularly as observed in the Beninese rural areas. This significantly affects popular engagement, which is clearly more limited in those areas, and feeds the denial. Emerging from the interviews are many reasons given for the denial of COVID-19 and vaccine refusal, including alleged environmental or genetic reasons and the so-called "neo-colonialism". In fact, the population believes that more than as a health emergency, this pandemic could be better described as a political alarmism with economic aims. This latter motivation finds its roots in the centuries of colonialism that still persist in the memory of the population so vividly; mainly that some people believe and spread conspiracy theories against the African population. One of the most interesting supported theories is that the low prevalence of the disease does not justify preventive prophylaxis compared to many other more common diseases with much higher mortality rates that have long deserved the search for a vaccine or cure. It would seem that a common belief is that the world wants to vaccinate Africa to protect itself, and that if the pandemic is not eradicated globally, there will always be the danger of its comeback, possibly in the form of even newer variants (two of the most recent ones were already firstly identified in Africa). However, on closer inspection it is clear from the interviews that local people do not want to fight a battle which is being disputed elsewhere, i.e., a disease that is not of concern in their own country. Even more so, local people are suspicious and angry about the underlying faulty mechanism behind it all (e.g., the COVID-19 vaccine discovery and testing in such short times), which did not consider as priority other epidemics or diseases mainly affecting Africa.

For these reasons, one of the most obvious lessons learnt from the pandemic is the urgent need to reflect once again on the *priorities*, that is, the ethical criteria that drive the progress of science, globally. Generally, the interviews highlighted the belief that third parties (international and/or local politicians, pharmaceutical companies, and "whites") for various reasons (economic, political, and experimental) are conspiring against the population (i.e., *zangbeto* metaphor). These fears reveal once again a clear gap between the people and the politicians, creating a situation where the populace fear the behavior of elected individuals who, in a democratic government, should represent them. This evident crisis of democracy motivates the attachment to popular beliefs concerning political charlatans, or those spread via social media, or by conventional wisdom.

As mentioned earlier, the most recurrent feeling interwoven with the interviews is that of fear. Fear has always been a philosophical, as well as a psychological concept on which many contemporary theorists have developed important reflections: from the heuristics of fear, theorized by Hans Jonas, for which profiling a terrifying future picture of the consequences of one's actions could favor the "principle of responsibility" (34), capable of promoting responsible actions to protect future generations, up to the reflections of the American philosopher Martha Nussbaum, a theorist of the social democracy based on skills and capabilities, who also modernly argues about the "monarchy of fear" (35), according to which fear is no longer part of an emotional dimension of the private sphere, but is part of a wider public sphere, affecting mass behaviors, just as it happened during the so-called "COVID-19 era".

In the case of Benin, the fear of oppression, linked to historical resentment towards the West, has led to the habit of identifying external culprits, thus fueling undemocratic attitudes. In this respect, Nussbaum argues that the wrath-blame mechanism is typical of fear. On closer inspection, however, fear is a primitive, narcissistic, and asocial emotion that is still causing more serious consequences in some countries than the spread of the virus itself. The fear of the vaccine is not a new thing for animist cultures. According to general belief, inoculation introduces pathogens that cause "something sick and evil", a situation which most would interpret in a spiritual sense (36). Currently, the traditional fear of the vaccine is intensified, evidenced by the refusal of the COVID-19 vaccine, with the implication that people might begin to reject vaccinations for other diseases too. Nonetheless, this should not lead us to think that the population rejects the COVID-19 vaccine for religious reasons, though it is a phenomenon common to several religions, which verified the compliance of vaccine serums with their religious dietary requirements. Islam and Judaism, for example, prohibit the intake of pork and its derivatives, while the Hindu religion prohibits the intake of cattle derivatives (37).

On the contrary, in Benin, the extraordinary tradition of traditional medicine, cloaked in its dual spiritual and phytotherapeutic nature, is readily available for searching causes and remedies related to the pandemic. Unfortunately, the government appears to distance itself from what represents very valuable local knowledge, even for primary healthcare. In fact, not only has the government refused to do an in-depth study of traditional medicines by assessing their effects and establishing their correct dosages, but also withdrew some herbal remedies that were either being used against COVID-19 or for strengthening the immune system from the open market (e.g., *Artemisia*). This disposition again shows the government's detachment from a confused and frightened population left to act alone and in secret. Certainly, in times of emergency, it is expensive and tedious to undertake studies parallel to official ones on alternative remedies to COVID-19, but it is also true that traditional medicine experts could support research and, at the same time, represent a reliable reference for the community to facilitate the engagement of the population (38). Notably, an attempt to officially appoint a group of traditional medicine experts to provide a solution to COVID-19 was made, but was unsuccessful due to a lack of organization and capacity for initiative.

Overall, as we have seen, there are many reasons behind the rejection of the COVID-19 vaccine. Among all, social engagement should be considered a priority. When considering feedback from the interviews, it would seem that social engagement poses a greater challenge than the much debated issue of the unfairness in the distribution of vaccines. Even if there were the availability of free and universal vaccination coverage, the popular rejection of vaccination would yet remain an obstacle to global health.

The aforementioned document of the Beninese government announcing the opening of the vaccination campaign states "Vaccination against COVID-19 is effective, safe, voluntary, and free. Get vaccinated to protect yourself and others".[16] Voluntariness corresponds to the ethical principle of *autonomy* that ascribes to the individual the balance between *risks* and *benefits*. According to the interviews, it would seem that the population is particularly frightened by the risks (including post-injection complications, scams made by governments, and the fear of dying). The benefit of vaccination, that is, individual and mass immunization, is underestimated because, from the personal point of view expressed by most interviewees, it is believed that other diseases are more widespread in those places (39) and there is no urgent need for this specific vaccine. The prospect of achieving global health should instead be mindful of historical heritage since the local belief is that the population is being used as "guinea pigs" to test health solutions for the sake of others.

Since people would be naturally more inclined to choose personal interests over collective ones, it is highly imperative that they are properly educated in making the right choices, especially as this improves the fight against possible pandemics and their behavior in times of emergency. Moreover, it is not possible to predict whether in the future the countries which are now relatively unaffected by the pandemic will witness an increase in infections, resulting in even more disastrous technical and medical consequences. This is why the Beninese government has not followed the risk-benefit balance, but rather, the *precautionary principle* in promoting mass vaccination of the population. However, considering the low record of population engagement in the fight against COVID-19, if the Beninese government wants to pursue the immunization goal, the question to ask is: should it give up the *autonomy principle*? From a theoretical point of view, it should be pointed out here that the *principle of autonomy*, which is at the basis of informed consent and dissent, is exercised when a competent individual, i.e., someone made clearly aware of the health framework, makes a decision in accordance with their moral perspective. However, in the present case of COVID-19, and taking infodemics into consideration, it would seem that the lucid and genuine self-determination of the individual has been seriously compromised. This is well linked to the ethical principle of *trust*, which is not built on authority, but on the relationship, involvement, and genuine inclusion of the population in scientific and political healthcare management. Trust should be built with clarity and honesty, and with sufficient explanations on any eventual limit or uncertainty of science (40). As of now, the emergence of rapid publication of information has called into question the accuracy of the information being circulated (41, 42). This phenomenon can be stopped by appealing first to scientists and researchers to be responsible and conscientious in their work, before even reaching out to those responsible for the circulation of information on communication networks and/or social media.

One of the first steps to clearer management of the pandemic would be to follow the infections more strictly with more capillary testing and digital tracking mechanisms (43). Many countries have followed this path, starting from the "immunity passport" (44) (which certifies whether a person has received the COVID-19 vaccine or has recently tested negative[17]). These passports permit holders to return to some of their normal activities, such as traveling more freely and returning to work (45). An immunity certification program could complement population or community-based strategies to ease restrictive measures, as well as secure societal and individual freedom and well-being.

Leaving aside the non-negligible issue of limited access to technologies (such as smartphones) and internet coverage, which could be the first obstacle to full deployment of the procedure, Benin has put in place an online portal hosting an app that can be used by vaccinated people to download and show a QR code as a proof of their status. Moreover, due to the limited number of people who accepted the vaccine, the government is prevented from limiting access to certain activities to only the vaccinated people, as well as from returning to severe containment measures such as lockdowns. Certainly, the immunity certification should never be used as the main strategy for reducing the effects of the COVID-19 pandemic, but could be used as a part of a plan that provides for a decrease in the number of people subject to highly restrictive measures (46, 47). As it is now, the immunity certification is an intermediate management strategy between mere recommendation and obligation, with related sanitation measures for whistleblowers – a balance between autonomy and public health.

The trust of the population must be built based on the recognition of local actors as community leaders, or even *tradithérapeutes* that can facilitate popular engagement in the vaccination program (48, 49). This is related to the ethical principle of *equal moral respect,* which is particularly interesting in relation to our study. This principle provides that "the interests of all be taken into equal account" and also requires "being sensitive to cultural diversity and plurality", which in turn requires a willingness to engage in dialogue and deliberation in terms of equal standing and recognition (…). In practice, this means ensuring that potential participants are empowered to reach their own decision regarding whether they would like to participate, ensuring that consent is sought in a culturally appropriate manner, and addressing participants' perspectives or concerns about the research, including information about how their data and samples will be handled, and so forth".[18] This could be achieved by an adequate study of local medicines and traditional therapeutic practices that can go hand in hand with the development of modern medicine.

This brings us back to the ethical principle of *equity*, which is not only related to a fair distribution of resources, but also to a consideration of specific popular needs: "for people to be treated equitably, they should be able to exert at least some influence over the decision-making process as well as the decision itself, i.e., procedural fairness".[19] Beyond this, including individuals and communities means maximizing the social value of science and politics (50). All points mentioned beforehand respond to the ethical principle of *solidarity* (51), which encompasses the respect of the pluralism of cultures and human rights. In the present case, human rights pertaining to health initiatives are individual and at the same time global, and go hand in hand with the right to enjoy the benefits of science and its results (in this case COVID-19 vaccines), all in respect of an individual's right to cultural identity. Furthermore, the latter point should be related to the right to having an adequate education, which puts a person in a position to clearly understand what is happening so as to exercise autonomous choices responsibly.

Conclusion

The world shares a collective responsibility in fighting the present pandemic; therefore, the reluctance of many local populations to accept the vaccine or their hesitancy towards tracking COVID-19 cases highlighted the urgent need for a study on social engagement in the fight against COVID-19. In particular, this paper shows the example of Benin, a Sub-Saharan country characterized by multiple endogen differences and a rich cultural tradition of animism and traditional medicine. With this interdisciplinary study, we explored the popular perspectives and beliefs on the pandemic from a sociological, ethical, political, and scientific point of view. In order to be able to use a multi-angled view and engage in multi-level discussions, an interdisciplinary framework was used. This framework, which makes use of multiple methodologies, allows schematization within a broad question and a number of issues, and provides tools to facilitate discussions at different levels. Moving away from a top-down approach and favoring a bottom-up one is considered an important way for multi-directional learning and for shaping an effective response informed by values.

The outputs emerging from the interviews, however, suggest a clear separation between the public and private sphere, and the distance of politics from the population, especially in peri-urban and rural areas, where negative perceptions and feelings of being ostracized prevail. The objective of this framework is to bring to public attention particularistic perspectives and how these suffer from the imposition of universal standards and procedures, i.e., in this case, the protocols of vaccine administration and tracking. The framework proposes flexible solutions to address these gaps (e.g., social mobilization strategy through the involvement of local medical expertise) because social engagement is vital to engage an effective immunization process. Current researchers should understand which modalities of public engagement are most effective. This study illustrates the urgent need for connecting clinical practice, public health, and social policy decision tables with broader community concerns, while also relying on bioethics. For the future, one of the lessons that should remain from the pandemic is that public perception should be considered a priority in the future management of healthcare emergencies, thus delivering the stamp of a "job well done" to posterity. One might argue that one of the limitations of this study is the limited sample size, which may not be representative of the whole Beninese populations. However, both the fact that our sample includes people with different cultural and socio-economic backgrounds, and our use of the aforementioned saturation threshold to capture all the possible themes, make this limitation negligible. Nonetheless, further comparative ethnography studies could focus on other populations of Sub-Saharan Africa to see whether similar beliefs and perceptions are shared within wider communities.

Notes

1 A. Maccaro, D. Piaggio, I. B. Oronti, M. Vignigbé, A. Gbokli, R. Hougnihin, L. Pecchia, *Social engagement in the fight against COVID-19 in the urban and peri-urban areas of Cotonou (Benin, Sub-Saharan Africa): acceptability of the vaccination and tracking program*, in «Frontiers in Medicine», 3 June 2022: www.frontiersin.org/articles/10.3389/fmed.2022.857890/full (https://doi.org/10.3389/fmed.2022.857890).

2 The Pfizer/BioNTech Comirnaty, 31 December 2020. The SII/COVISHIELD and AstraZeneca/AZD1222 vaccines, 16 February 2021. The Janssen/Ad26.COV 2.S developed by Johnson & Johnson, 12 March 2021. The Moderna COVID-19 vaccine

(mRNA 1273), 30 April 2021.The Sinopharm COVID-19 vaccine, 7 May 2021. The Sinovac-CoronaVac, 1 June 2021. The Bharat Biotech BBV152 COVAXIN vaccine, 3 November 2021.

3 https://covid19.who.int/
4 https://covid19.who.int/table
5 https://ourworldindata.org/covid-vaccinations
6 www.who.int/initiatives/act-accelerator/covax
7 www.who.int/news/item/23-09-2020-managing-the-covid-19-infodemic-promoting-healthy-behaviours-and-mitigating-the-harm-from-misinformation-and-disinformation
8 https://unsdg.un.org/latest/stories/benin-national-covid-19-vaccination-campaign-has-launched
9 www.gouv.bj/coronavirus/#documents
10 https://covid19.who.int/region/afro/country/bj
11 The study received full ethical approval from the University of Abomey-Calavi Comitè d'Ethique de la Recherche CER-ISBA (N° 131).
12 www.afro.who.int/news/benin-goes-digital-offensive-against-covid-19
13 The *zangbeto* is the composition of two phonemes: "zan" meaning "night" and "gbeto" meaning "hunter", i.e., night hunter. The *zangbeto* is, from a social point of view, the traditional night watchman and a secret society of masks in southern Benin. When he goes out, the *zangbeto*, being masked, is accompanied by *zanvi* (followers of the *zangbeto*) and if he is caught by a mouse trap, it means that he is a victim of a trap. Otherwise, the people are victims of a conspiracy.
14 *Sapkata* is a deity of southern Benin, of Yoruba origin, reputed to be one of the dominant figures of the Vodun pantheon. It is considered as the regent on earth and the God of eruptive diseases. It is usually the beneficiary of expiatory rites in cases of epidemic.
15 *Soudjo* and *Tokplopklo* are expiatory rites literally known as baths of purification for the community life to ostracize the evil from the socio-linguistic group
16 www.gouv.bj/coronavirus/#documents
17 Although this second option gives very limited freedom as the obtained pass lasts only a few days.
18 Dawson, Angus, et al. "Key ethical concepts and their application to COVID-19 research." *Public Health Ethics* 13.2 (2020): 127-132, 128-129.
19 Ivi, p. 129.

References

(1) Pecchia L, Piaggio D, Maccaro A, Formisano C, Iadanza E. The inadequacy of regulatory frameworks in time of crisis and in low-resource settings: Personal protective equipment and COVID-19. *Health and Technology* [Internet]. 2020 November [cited 2022 January 17];10(6):1375–1383. Available from: https://doi.org/10.1007/s12553-020-00429-2

(2) Maccaro A, Piaggio D, Pagliara S, Pecchia L. The role of ethics in science: A systematic literature review from the first wave of COVID-19. *Heal Technol [Internet]*. 2021 September [cited 2022 January 17];11(5):1063–1071. Available from: https://link.springer.com/10.1007/s12553-021-00570-6

(3) Maccaro A, Piaggio D, Dodaro CA, Pecchia L. Biomedical engineering and ethics: Reflections on medical devices and PPE during the first wave of COVID-19. *BMC Medical Ethics* [Internet]. 2021 December [cited 2022 January 17];22(1):130. Available from: https://bmcmedethics.biomedcentral.com/articles/10.1186/s12910-021-00697-1
(4) European Centre for Disease Prevention and Control. SARS-CoV-2 variants of concern as of 13 January 2022 [Internet]. [cited 2022 January 17]. Available from: www.ecdc.europa.eu/en/covid-19/variants-concern
(5) World Health Organization. Tracking SARS-CoV-2 variants [Internet]. [cited 2022 January 17]. Available from: www.who.int/emergencies/what-we-do/tracking-SARS-CoV-2-variants
(6) Le TT, Andreadakis Z, Kumar A, Roman RG, Tollefsen S, Saville M, et al. The COVID-19 vaccine development landscape. *Nature Reviews Drug Discovery* [Internet]. 2020 [cited 2022 January 18];19(5):305–306. Available from: www1.cgmh.org.tw/library_s/2020/paper/P-0428 The COVID-19 vaccine development landscape.pdf
(7) Dawson A, Emanuel EJ, Parker M, Smith MJ, Voo TC. Key ethical concepts and their application to COVID-19 research. *Public Health Ethics* [Internet]. 2020 May [cited 2022 January 17];phaa017. Available from: www.ncbi.nlm.nih.gov/pmc/articles/PMC7239092/
(8) Faden R, Saad O, Kochhar S, Kaslow D, Pallas S, Olayinka F, et al. WHO SAGE values framework for the allocation and prioritization of COVID-19 vaccination. 2020 September 14 [cited 2022 January 18]. Available from: https://apps.who.int/iris/handle/10665/334299
(9) WHO. WHO SAGE roadmap for prioritizing uses of COVID-19 vaccines in the context of limited supply: An approach to inform planning and subsequent recommendations based on epidemiological setting and vaccine supply scenarios. 2021;[October 2020]:1–24. Available from: https://apps.who.int/iris/handle/10665/342917
(10) Troiano G, Nardi A. Vaccine hesitancy in the era of COVID-19. *Public Health* [Internet]. 2021 May [cited 2022 January 17];194:245–251. Available from: www.sciencedirect.com/science/article/pii/S0033350621000834
(11) Machingaidze S, Wiysonge CS. Understanding COVID-19 vaccine hesitancy. *Nature Medicine* [Internet]. 2021 August [cited 2022 January 17];27(8):1338–1339. Available from: www.nature.com/articles/s41591-021-01459-7
(12) Solís Arce JS, Warren SS, Meriggi NF, Scacco A, McMurry N, Voors M, et al. COVID-19 vaccine acceptance and hesitancy in low and middle income countries, and implications for messaging [Internet]. 2021 March [cited 2022 January 17]. Available from: http://medrxiv.org/lookup/doi/10.1101/2021.03.11.21253419
(13) van der Linden S, Dixon G, Clarke C, Cook J. Inoculating against COVID-19 vaccine misinformation. *eClinicalMedicine* [Internet]. 2021 March [cited 2022 January 17];33:100772. Available from: https://linkinghub.elsevier.com/retrieve/pii/S2589537021000523
(14) Infodemic [Internet]. [cited 2022 January 17]. Available from: www.who.int/westernpacific/health-topics/infodemic
(15) Zarocostas J. How to fight an infodemic. *Lancet* [Internet]. 2020 February [cited 2022 January 17];395(10225):676. Available from: https://linkinghub.elsevier.com/retrieve/pii/S014067362030461X

(16) WHO. An ad hoc WHO technical consultation managing the COVID-19 infodemic: Call for action [Internet]. *Infodemic Management*. 2020 [cited 2022 January 18]. Available from: https://apps.who.int/iris/bitstream/handle/10665/334287/9789240010314-eng.pdf?sequence=1&isAllowed=y

(17) UNESCO. Disinfodemic: Deciphering COVID-19 disinformation [Internet]. 2021 [cited 2022 January 18]. Available from: https://en.unesco.org/covid19/disinfodemic/brief1

(18) Bhopal S, Nielsen M. Vaccine hesitancy in low- and middle-income countries: Potential implications for the COVID-19 response. *Archives of Disease in Childhood* [Internet]. 2021 February [cited 2022 January 17];106(2):113–114. Available from: https://adc.bmj.com/lookup/doi/10.1136/archdischild-2020-318988

(19) WHO. Microsoft Power BI: Africa COVID-19 Vaccination Update [Internet]. 2021 [cited 2022 January 19]. Available from: https://app.powerbi.com/view?r=eyJrIjoiZmMwYzJiZjktZWE5MS00ZjlkLThkZjItZjlkNWNiZjc5OTc1IiwidCI6ImY2MTBjMGI3LWJkMjQtNGIzOS04MTBiLTNkYzI4MGFmYjU5MCIsImMiOjh9

(20) Ackah BBB, Woo M, Ukah UV, Fazal ZA, Stallwood L, Okpani A, et al. COVID-19 vaccine hesitancy in Africa: A scoping review [Internet]. 2021 August [cited 2022 January 17]. Available from: www.researchsquare.com/article/rs-759005/v1

(21) Bono SA, de Moura Villela E, Siau CS, Chen WS, Pengpid S, Hasan MT, et al. Factors affecting COVID-19 vaccine acceptance: An international survey among low- and middle-income countries. *Vaccines* [Internet]. 2021 May [cited 2022 January 17];9(5):515. Available from: www.mdpi.com/2076-393X/9/5/515

(22) Arakpogun EO, Elsahn Z, Prime KS, Gerli P, Olan F. Digital contact-tracing and pandemics: Institutional and technological preparedness in Africa. *World Dev [Internet]*. 2020 December [cited 2022 January 17];136:105105. Available from: www.ncbi.nlm.nih.gov/pmc/articles/PMC7396131/

(23) Adebisi YA, Oke GI, Ademola PS, Chinemelum IG, Ogunkola IO, Lucero-Prisno III DE. SARS-CoV-2 diagnostic testing in Africa: Needs and challenges. *Pan African Medical Journal* [Internet]. 2020 April [cited 2022 January 17];35(Supp 2). Available fromwww.panafrican-med-journal.com/content/series/35/2/4/full

(24) Onebunne JI, Okechukwu S, Okoye BC. African ecophilosophy and COVID-19. In Kanu IA, editor. *Trends and advances in education and management in Africa* [Internet]: 2021; [cited 2022 January 19]. pp. 199–219. Available from: www.researchgate.net/publication/356377633

(25) Onebunne J, Okechukwu S, Okoye B. COVID-19 and African environmental/herbal based remedies. In: Kanu IA, editor. *African eco-philosophy: Cosmology, consciousness and the environment*: 2021, pp. 275–296.

(26) Akubor EO. The African forests in the face of the Coronavirus (COVID 19) pandemic. In Kanu IA, editor. *African ecological spirituality: Perspectives in anthroposophy and environmentalism a hybrid of approaches*: 2021, pp. 151–205.

(27) Piaggio D, Castaldo R, Cinelli M, Cinelli S, Maccaro A, Pecchia L. A framework for designing medical devices resilient to low-resource settings. *Global Health*. 2021 Jun 22;17(1):64.

(28) Saunders B, Sim J, Kingstone T, Baker S, Waterfield J, Bartlam B, et al. Saturation in qualitative research: Exploring its conceptualization and operationalization. *Qual Quant* [Internet]. 2018 July 1 [cited 2022 April 23];52(4):1893. Available from: www.ncbi.nlm.nih.gov/pmc/articles/PMC5993836/

(29) Baker SE, Edwards R. How many qualitative interviews is enough? Expert voices and early career reflections on sampling and cases in qualitative research.

(30) WHO. WHO's Infection prevention & control department [Internet]. 2022 [cited 2022 January 19]. Available from: www.who.int/teams/integrated-health-services/infection-prevention-control
(31) Nordling L. Unproven herbal remedy against COVID-19 could fuel drug-resistant malaria, scientists warn. *Science* [Internet]. 2020 [cited 2022 January 17]; Available from: https://doi.org/10.1126/science.abc6665
(32) Di Pietro L, Piaggio D, Oronti I, Maccaro A, Houessouvo RC, Medenou D, et al. A framework for assessing healthcare facilities in low-resource settings: Field studies in Benin and Uganda. *Journal of Medical and Biological Engineering* [Internet]. 2020 August [cited 2022 January 17];40(4):526–534. Available from: https://link.springer.com/10.1007/s40846-020-00546-3
(33) Piaggio D, Medenou D, Houessouvo RC, Pecchia L. Donation of medical devices in low-income countries: Preliminary results from field studies. In Badnjevic A, Škrbić R, Gurbeta Pokvić L, editors. *International Conference on Medical and Biological Engineering*: Springer; 2019 [cited 2022 January 17] pp. 423–427. Available from: http://link.springer.com/10.1007/978-3-030-17971-7_64
(34) Jonas H. *The imperative of responsibility: In search of an ethics for the technological age* [Internet]. University of Chicago Press; 1984 [cited 2022 January 19]. Available from: https://philpapers.org/rec/JONTIO-4
(35) Nussbaum MC. The monarchy of fear: A philosopher looks at our political crisis [Internet]. *Simon & Schuster*. 2019 [cited 2022 January 19]. Available from: https://doi.org/10.1057/s41296-018-00305-9
(36) Fourn L, Haddad S, Fournier P, Gansey R. Determinants of parents' reticence toward vaccination in urban areas in Benin (West Africa). *BMC International Health and Human Rights* [Internet]. 2009 October [cited 2022 January 18];9(S1):S14. Available from: https://bmcinthealthhumrights.biomedcentral.com/articles/10.1186/1472-698X-9-S1-S14
(37) Williams JTB, Fisher MP, Bayliss EA, Morris MA, O'Leary ST. Clergy attitudes toward vaccines and vaccine advocacy: A qualitative study. *Human Vaccines & Immunotherapeutics* [Internet]. 2020 March [cited 2022 January 18];16(11):2800–2808. Available from: www.ncbi.nlm.nih.gov/pmc/articles/PMC7734036/
(38) Maccaro A. Beninese therapeutic pluralism. Historical roots and bioethical issues. *Journal of Theoretical Applied Vascular Research* [Internet]. 2018 [cited 2022 January 18];3(1):19–24. Available from: https://core.ac.uk/reader/195735400
(39) Smith MJ, Upshur REG. Learning lessons from COVID-19 requires recognizing moral failures. *Journal of Bioethical Inquiry* [Internet]. 2020 December [cited 2022 January 18];17(4):563–566. Available from: https://link.springer.com/10.1007/s11673-020-10019-6
(40) Giubilini A, Douglas T, Savulescu J. The moral obligation to be vaccinated: Utilitarianism, contractualism, and collectives easy rescue. *Medicine, Health Care and Philosophy* [Internet]. 2018 December [cited 2022 January 18];21(4):547–560. Available from: http://link.springer.com/10.1007/s11019-018-9829-y
(41) Smith MJ, Upshur REG, Emanuel EJ. Publication ethics during public health emergencies such as the COVID-19 pandemic. *American Journal of Public Health* [Internet]. 2020 July [cited 2022 January 18];110(7):947–948. Available from: https://ajph.aphapublications.org/doi/full/10.2105/AJPH.2020.305686
(42) Smith MJ, Ahmad A, Arawi T, Dawson A, Emanuel EJ, Garani-Papadatos T, et al. Top five ethical lessons of COVID-19 that the world must learn. *Wellcome Open Research*

[Internet]. 2021 January [cited 2022 January 18];6:17. Available from: www.ncbi.nlm.nih.gov/pmc/articles/PMC7919608/

(43) WHO. Ethical considerations to guide the use of digital proximity tracking technologies for COVID-19 contact tracing interim guidance [Internet]. 2020 May [cited 2022 January 19]. Available from: https://apps.who.int/iris/bitstream/handle/10665/332200/WHO-2019-nCoV-Ethics_Contact_tracing_apps-2020.1-eng.pdf

(44) Persad G, Emanuel EJ. The ethics of COVID-19 immunity-based licenses ("Immunity Passports"). *JAMA* [Internet]. 2020 June [cited 2022 January 18];323(22):2241. Available from: https://jamanetwork.com/journals/jama/fullarticle/2765836

(45) Brown RCH, Savulescu J, Williams B, Wilkinson D. Passport to freedom? Immunity passports for COVID-19. *Journal of Medical Ethics* [Internet]. 2020 October [cited 2022 Januarry 18];46(10):652–659. Available from: https://jme.bmj.com/lookup/doi/10.1136/medethics-2020-106365

(46) WHO. "Immunity passports" in the context of COVID-19 [Internet]. Geneva; 2020 [cited 2022 January 18]. Available from: www.who.int/news-room/commentaries/detail/immunity-passports-in-the-context-of-covid-19

(47) Voo TC, Reis AA, Thomé B, Ho CWL, Tam CC, Kelly-Cirino C, et al. Immunity certification for COVID-19: Ethical considerations. *Bulletin of the World Health Organization* [Internet]. 2021 February [cited 2022 January 18];99(2):155–161. Available from: www.ncbi.nlm.nih.gov/pmc/articles/PMC7856365/

(48) Faye SLB, Diouf W, Cisse PN, Quach A, Sylla MM, Koita MM, et al. Engager (avec) les communautés dans un essai vaccinal en contexte post-Ebola (Guinée Conakry): un modèle basé sur les « champions »: *Sci & Actions Sociales* [Internet]. 2018 November [cited 2022 January 18];10(2):112–141. Available from: https://cairn.info/revue-sciences-et-actions-sociales-2018-2-page-112.htm?ref=doi

(49) Sarradon A, Egrot M, Blanc M-A, Faure M. Approche anthropologique des déterminants de l'observance dans le traitement de l'hypertension artérielle. 2008;39(1):3–12.

(50) Jalloh MF, Wilhelm E, Abad N, Prybylski D. Mobilize to vaccinate: Lessons learned from social mobilization for immunization in low and middle-income countries. *Human Vaccines & Immunotherapeutics* [Internet]. 2020 May [cited 2022 January 18];16(5):1208–1214. Available from: www.tandfonline.com/doi/full/10.1080/21645515.2019.1661206

(51) Bruce L. Global solidarity in COVID-19 vaccine distribution. *BioLaw Journal – Rivista di BioDiritto.* 2021;1:5.

A Framework for Designing Medical Devices Resilient to Low-Resource Settings[1]

Background

According to the World Bank and the World Health Organization (WHO), at least half of the world population had no access to essential health services in 2017 (1). In fact, Arasaratnam et al. (2) affirmed that only 13% of the global population accounts for 76% of the global MD use, suggesting inequitable healthcare access in favor of higher-resource settings. The latter also preside over the medical device (MD) market, 80% of which is ruled by the United States, Europe, and Japan (3). These countries set and represent a *de facto* standard and can rely on medical settings that conform to international standards of quality, which should not be taken for granted in low-resource settings (LRSs). This can be one of the reasons why in lower-income settings up to 70-90% of donated MDs (2, 4, 5) are non-operational or broken. Indeed, this often relates to the majority of available devices, for example 80% of the MDs of Sub-Saharan Africa are donated (6).

Low-resource settings, in particular, are characterized by generally poor environmental and operational conditions. In fact, healthcare is typically delivered in locations which are not fit for purpose (e.g., dust, vermin, humidity, high temperatures, etc.) (6, 7). The literature highlights the main challenges for healthcare provision in LRSs to be: lack of funding, insufficient essential MDs (7–10), lack of specialized doctors (7–11), lack of biomedical engineers and technicians (who can maintain or repair MDs), and deficiency of spare parts/consumables (7–10). Further, the situation and needs can change significantly between city centers and the periphery. In 2007, Malkin (12) identified that capital cost, lack of spare parts and lack of consumables were the three main barriers to the introduction of healthcare technology in hospitals in LRSs. All these factors negatively influence the efficacy, safety and durability of MDs.

In the 2010s, some studies started relating this situation to a design problem. For instance, Aranda et al. (13), similarly to Kortum et al. (14), stated that healthcare facilities are in poor conditions in terms of basic infrastructure due to designers' failure. Aranda et al. (11) also asserted that these issues could be efficiently tackled with a well-planned, user-driven, and contextualized design. Similarly, to manufacture safe, efficient, and effective MDs, it is essential to define both the user requirements and

DOI: 10.1201/9781003436348-8

the context/environment of use, which in some cases are not considered as part of the design process (15).

In the literature, there have been attempts to identify the crucial criteria for the design of MDs resilient to LRSs (2, 5, 13, 16). Arasaratnam et al. (2) gave an overview of the past trend, better known as "glocalisation". This term, not to be confused with the frequently misused term "globalization", arises from the combination of the latter with the term "localization". It is used to describe a phenomenon of contextualization of some products or services, universally distributed but adjusted/modified to accommodate the need of the user or consumer of local markets. Glocalisation was a typical trend of the industry, which removed some features from high-tech products that were designed for more developed countries to make them accessible to LRSs. Nowadays, it is widely recognized that this trend is not sufficient to adjust medical technologies to LRSs. Thus, the new trend is so-called "frugal innovation" (17). The study by Aranda et al. (13), relying on experts' views, identified some essential criteria in the durability, robustness, and cost of ownership over time and simplicity of use. The same criteria were pointed out by Nimunkar et al. (16) along with accuracy, reliability, size, weight, materials, power requirements, ease of manufacturing, language barriers, availability of facilities, and population dynamics. The latter also agreed with the widely shared belief that special considerations should be made when designing MDs for developing countries. Similar results were obtained by Gauthier et al. (5) through a literature review and an analysis of state-of-the-art MDs designed for LRSs. Beyond the already mentioned criteria, Gauthier et al. (5) also suggested further criteria such as appropriateness, functionality, spare parts, personnel, management, and public policy. Moreover, the authors also asserted that the primary focus should be on affordability, durability, and materials (5).

Interestingly, in the nineties Adler et al. (18) presented and described the application of some military standards to the design of MDs for military use. This did not seem to receive as much attention as the current (2021) trend followed by the Information Technology (IT) industry, which is adopting the Ingress Protection (IP) Code (International Electrotechnical Commission [IEC] standard 60529) and military standards (specifically, MIL-STD-810H) to design and manufacture "ruggedised" devices, which are sturdier and more resilient to harsh working conditions. Nonetheless, there is no overall agreement on a framework including a set of criteria that supports the design of MDs for LRSs, outlining the key considerations and desirable properties.

The aim of this study is to contribute to bridging this research gap by systematically developing a general framework that could guide the design of MDs for LRSs. In particular, this study aims to identify, classify, and weight the essential criteria to be considered while designing MDs resilient to LRSs. The framework was initiated starting from the MDs presented in the WHO compendium of innovative health technologies for LRSs (19).

Our goal was pursued inductively, i.e., by inferring the general set of essential criteria from observations/data. A triangulation of qualitative and quantitative methods was used. In fact, as Pope et al. affirmed (20), the former can complement the latter when used as an essential preliminary step to quantitative research and can also be supplementary information to further validate the quantitative results. Such methods, which will be thoroughly described in the paper, include: focus groups, field studies, content analysis for drafting and validating a set of essential criteria, pilot tests, semi-structured interviews, Delphi survey, and multi-criteria decision analysis (MCDA) for developing and validating a questionnaire, and analyzing the results. As a result, the final version of the questionnaire (see Additional File 1) was issued to a pool of experts to complete. Consequently, a set of essential criteria was obtained and used to develop a framework, i.e., a multi-criteria decision system for designing MDs resilient to LRSs.

This work could lay the basis for the creation of novel and contextualized international standards, which could empower LMICs, enabling them to enforce import restrictions according to the World Trade Organization Agreement on Trade-Related Aspects of Intellectual Property Rights (WTO-TRIPS), and creating "an indigenous medical equipment manufacturing capability" (12). Thus, this would force manufacturers to design and manufacture products in line with the new consensus standard to give them the opportunity to market their products in LMICs.

Methods

The inductive method, which implies the generation of theories starting from numerous observations (21), was used to develop the set of essential criteria. In particular, a triangulation of qualitative and quantitative methods was used to develop a framework of criteria for the design of MDs resilient to LRSs and for its validation through the convergence of information (20, 22). Figure 2.3 summarizes the project phases along with their main objectives, methods, people involved, and outcomes. In particular, purposive sampling (20, 23) was used for the methods that required sampling. The latter is a non-random technique that consists of the researcher deliberately choosing participants, based on the information that they can provide relying on their knowledge or experience (24).

Criteria Identification

The identification, classification, and weighting of the fundamental criteria for the design of MDs for low-income settings were achieved with a series of nested closed loops involving relevant scholars and experts (see Figure 2.4) from five continents. All the steps presented in Figures 2.3 and 2.4 are interdependent.

After analyzing the criteria for the design of medical locations and MDs, essential criteria for the resilient design for LRSs were identified by systematically reviewing the existing scientific literature. These initial criteria and the rationale of this study were presented at the Africa Health Conference (Johannesburg, South Africa, June

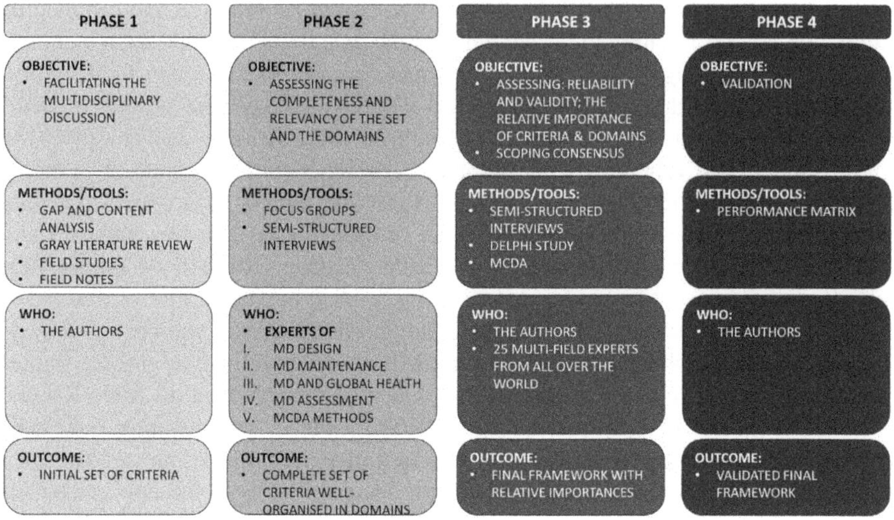

FIGURE 2.3
A summary of the phases, their objectives, methods, people involved and outcomes.

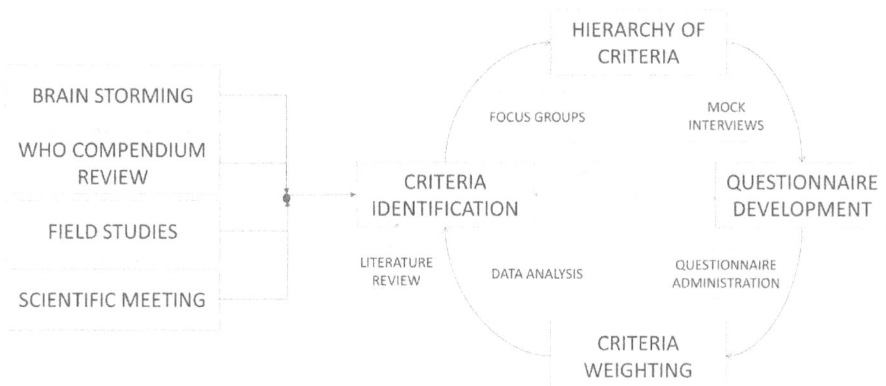

FIGURE 2.4
Block diagram of the study process and methods used.

2016) and discussed in a focus group with biomedical and clinical engineers from 15 Sub-Saharan Africa countries.

In November 2016, the essential criteria were reviewed during a three-day meeting organized in Warwick in collaboration with the International Federation for Medical and Biological Engineering's (IFMBE) Health Technology Assessment Division (HTAD) (http://htad.ifmbe.org/) working group for the preparation of guidelines for the Health Technology Assessment of MDs (25). During the meeting, 14 experts from eight nations and one WHO representative participated in a focus group to analyze the interdependencies among working environments and MDs safety and effectiveness (15). During this focus group, the essential criteria were enriched with experts' feedback and a grey literature review on LRSs, design of MDs, and medical locations. Particular attention was given to the WHO compendium of innovative health technologies for LRSs (19). Two authors (DP, SC) analyzed the compendium and reported the MDs that are supposedly resilient to harsh environments. The recurrent design criteria were noted, narrowed down, and finally enriched with the experts' views.

Hierarchy of Criteria

The identified criteria were clustered in meaningful domains defining a hierarchy framework. The hierarchy was piloted in Africa with several field studies, each followed by an additional focus group. In May 2017 and January 2018, two field analyses were carried out in Benin performing visual inspections and testing of MDs and medical locations (6) to evaluate the relevance and the suitability of the essential criteria set. The former field study was followed by a focus group with experts from Sub-Saharan Africa during the Africa Health Conference (Johannesburg, South Africa, June 2017). The latter field study was followed by a focus group held in Addis Ababa, Ethiopia, during the kick-off meeting of the IFMBE Working Group on biomedical engineering (BME) in Africa, in February 2018 (26). Additionally, in 2019 two extra field studies were carried out in Benin and Uganda (7).

Survey

Once the set of essential criteria was consolidated, one investigator (DP) created the first draft of the questionnaire to confirm the set of criteria, explore consensus among scholars and domain experts, and weight criteria with relative importance.

The survey was pilot-tested in two stages. Firstly, it was administered to the participants of the first focus group in Warwick for possible disambiguation and correction of content and language. Secondly, it was tested with in-person semi-structured mock interviews at the International Union for Physical and Engineering Sciences in Medicine (IUPESM) World Congress (Prague, June 2018), recruiting scholars not involved in the previous steps. In particular, during the IUPESM Congress, one of the authors (DP) conducted semi-structured interviews and collected the answers of biomedical and clinical engineering experts, whose feedback helped us refine the questionnaire and enrich it with one additional criterion. Eventually, another focus group was held to review the criteria and to create the final version of the questionnaire. The final questionnaire was sent with a bilingual email invitation, linking to two online questionnaires, one in English and one in French, to improve the reachability of African experts from francophone countries. One reminder was sent after one month to the panelists who did not reply.

Due to our extensive network of institutions, responders were identified amongst the ones collaborating with the IFMBE's Health Technology Assessment and Clinical

Engineering Divisions and speakers selected by the WHO for the sessions on MD design and management in LMICs of the Fourth WHO Global Forum on Medical Devices. Finally, authors of relevant scientific papers were invited, which were independent of the above institutions. Responders were recommended based on their experience with health technologies, including design, development, testing, implementation, and maintenance of MDs. The study received full ethical approval by the University of Warwick Biomedical and Scientific Research Ethics Committee (BSREC) (REGO-2018-2283).

Delphi Survey Design

Named after the famous oracle at Delphi, the Delphi study is one of the several consensus methods available (27). It is an iterative multistage process that aims to elicit and reinforce experts' opinions via group consensus with a series of structured questionnaires, which the experts fill in. This can be conducted in two main ways. The questionnaire can be anonymous, i.e., the respondents' identity is private and they cannot see another respondent's answers; or it can be blinded, i.e., the respondents' identity is only known by the investigators and they are still "blind" to others' answers. This allows every participant to freely express their own opinions, minimizes the "bandwagon effect" (i.e., the phenomenon in which people tend to do something just because other people do it, regardless of their own beliefs or opinions (28)) and prevents the authority or reputation of some participant from dominating others (29). In our study, we conducted a blind online Delphi survey, which allowed us to reach expert panelists all around the world. Our study aimed to obtain consensus merging the panelists' different opinions through group dynamics, rather than to achieve statistical power. Therefore, *a priori* power analysis to calculate the minimum sample number of panelists was not evaluated for this study. However, we referred to the existing literature that suggests that a panel should include ten to 18 experts (27).

Delphi Process

The questionnaire was composed of 42 questions, organized in nine sections: one regarding the responder's professional experience, seven weighting the importance of design criteria clustered in different domains, and one to weight the relative importance of such domains. Each section closed with an open question to gather suggestions about new criteria, if any. All the acronyms of the questionnaire were duly explained. Definitions were given for some of the criteria. In fact, given the respondents are experienced in the domain of MD design with a special focus on LRSs, we assumed that well-established concepts in this domain needed no further explanation (i.e. maintenance frequency, maintenance cost, the need for consumables or spare parts, etc.). Conversely, for the criteria that were not deemed self-explanatory, examples/definitions (please see Additional File 1) were added in brackets in the questions of the questionnaire as in the following examples:

For "reliance on medical location air", examples were given such as "filtering, temperature, humidity, etc."

For "end users' background", examples were given such as "doctors, nurses, biomedical engineering technicians"

For "ease of use", examples were given such as "the medical device requires a highly specialized personnel, any operator can use it, or it is so intuitive that anyone can use it" etc.)

The questions regarding the design criteria were asked with a repetitive formula, namely "*What is the importance of considering* "CRITERION 1" *during the design of MD which will be used in low-resource settings?*" Such questions had six possible answers, converted to an integer number based on a five-step Likert-type scale, i.e., "Very low" (1), "Low" (2), "Middle" (3), "High" (4), "Very high" (5), "Do not know" (" "). A new criterion would be included in a subsequent Delphi round if proposed by at least 30% of the respondents (25). Regarding the iteration stopping criteria, we decided not to iterate the questionnaire if all of the criteria and domain reached consensus defined as follows. For all the criteria, the first and third quartiles and their relative distances 'Δ', i.e., the interquartile ranges, were calculated in order to rank the criteria consensus with a "Sufficient" range ($2 \geq \Delta \geq 1$), "Fair" range ($1 \geq \Delta > 0$) or a "Full consensus" range ($\Delta = 0$).

Data Analysis

Reliability

When performing a survey, it is crucial to assess the reliability of the responses. Low reliability negatively affects the validity of the results (30). Reliability measures how the criteria are part of the same domain and are, thus, relevantly measuring the same overarching concept. This is fundamental since experienced experts may also provide inconsistent answers due to tiredness or distraction. Following the procedure previously employed in other studies (31, 32), reliability was estimated through ordinal alpha and the use of polychoric matrices. The latter are optimum tools for analyzing the reliability in case of ordinal data (such as Likert-scale based questionnaires) because Cronbach alpha is based on the assumption of continuous data (33) and could lead to underestimation (34). Ordinal alpha scale, in particular, ranges from 0 to 1, with values over 0.7 being "Acceptable" (31).

Polychoric correlation matrices and the consequent ordinal alphas were calculated in R (version 3.5.1) with the functions "polychoric()" and "alpha()" for each domain.

As suggested in (35), once the ordinal alphas were calculated, internal consistency was rated according to the values reported in Table 2.2. Then, the alpha drop was also calculated for the domains that showed low internal consistency. The latter is a measure of how the alpha would change if a criterion of the domain was dropped. This helps identify the problematic elements of the questionnaire, if any.

TABLE 2.2

The values of alpha and their interpretation.

Alpha	Internal Consistency
$\alpha \geq 0.9$	Excellent
$0.8 \leq \alpha < 0.9$	Good
$0.7 \leq \alpha < 0.8$	Acceptable
$0.6 \leq \alpha < 0.7$	Questionable
$0.5 \leq \alpha < 0.6$	Poor
$\alpha \leq 0.5$	Unacceptable

Validity

Validity is an essential component of survey evaluation; in fact, it tests whether the scale is measuring what it is intended to measure (31). In this case, validity was tested during the pilot test, in which interviewees were asked for additional comments and also during the Delphi survey. Each domain contained a final question asking to suggest criteria, their ranking, and possible comments. New criteria would be added if suggested by at least 30% of the panelists.

Ranking

In order to find the final importance of each criterion derived from the answers to the questionnaire, the relative index was calculated. The Relative Index (RI) is defined by the following equation (31):

$$RI = \sum_{i=1}^{5} \frac{w_i f_i}{N}$$

Where w_i is the weighting factor calculated by dividing the rating score by the highest score (i.e., 5), f_i is the frequency of the responses and N is the total number of responses. Once the relative indices were calculated, the criteria were sorted in decreasing order and divided into subclasses according to their importance, namely Fundamental, Important, and Relevant.

Correlations Among Criteria

Correlations among criteria in the same domain were assessed with Goodman and Kruskal's gamma, which was calculated with the following equation (31):

$$\gamma = \frac{\text{SOP} - \text{IOP}}{\text{SOP} + \text{IOP}}$$

Where *SOP* stands for "Same Order Pair" and *IOP* for "Inverse Order Pair". Gamma (γ) ranges from -1 to 1 and in this case values greater than 0.5 or lower than -0.5 were considered to highlight a strong correlation, whose significance depends on the p-value. In this case, we selected a p-value of 0.05.

Validation and Use of the Framework

Once identified, the relevant criteria were combined into a framework that could be used to help inform the design of MDs resilient to LRSs.

Subsequently, a protocol was developed to test this framework of criteria with a subset of MDs and to show its possible use. All the criteria were reported and divided into two sets: discriminatory and non-discriminatory criteria. Three criteria were added *a posteriori* of our field studies, under the "Reliance on external factors" domain.

A total of eight MDs was randomly extracted from the WHO compendia of innovative health technologies for LRSs, four from the 2014 version (19) and four from the 2017 version (36).

After analyzing the selected devices, information regarding each of the criteria was used to score them with the help of a performance matrix (similarly to the concept of Pugh matrix (37)). The criteria from the cost and lifetime domains, i.e., non-discriminatory criteria, were excluded from this analysis because they cannot be scored qualitatively. A qualitative three point-based measurement scale was used. Adopting the well-known traffic light color code, red was used for the worst cases, yellow for the intermediate performances, and green for the best cases. Consequently, two trends were evaluated:

1. Which criteria feature mainly green cells (i.e., a good assessment) irrespective of the device? In this case, only the criteria reaching a good assessment in at least five out of the eight MDs were selected.
2. Which of the selected MDs resulted efficient (i.e., criteria prevalently coded as green), which averagely efficient (i.e., criteria prevalently coded as yellow) and which not really efficient (i.e., criteria prevalently coded as red)?

Results

Focus group and pilot. The final version of the hierarchy of essential criteria for the design of MDs resilient to LRS working conditions resulted from the second focus groups and brainstorming session. Specifically, the criteria were grouped into seven main domains as follows:

1. User type
2. Health technology management (HTM)
3. Design
4. Reliance on external factors
5. Material
6. Cost
7. Lifetime

Moreover, the piloting phase led to the addition of one criterion, namely "the capability of the user to understand the technical and clinical impact of the technology".

Delphi Survey

Characteristics of Panelists

We invited 56 professionals, with a background spanning from biomedical engineering to clinical engineering and public health, to participate in the final survey, and an overall 25/56 (44.6%) answered the questionnaire. The 25 panelists represented 19 countries, 7/25 (28%) of them represented Europe, 12/25 (48%) Africa, 2/25 (8%) North America, 2/25 (8%) East Asia, 1/25 (4%) Central America, and 1/25 (4%) South America. As far as Africa is concerned, the involved countries were the Democratic Republic of Congo, Ethiopia, Tanzania, Mozambique, Benin, South Africa, Ghana, Kenya, and Burundi. The fields of expertise of the respondents are broad and summarized Table 2.3. Since some respondents had more than one area of expertise, the different areas of expertise were counted as separate entries. The

TABLE 2.3

A summary of the areas of expertise of the panelists. The numbers are out of 29 respondents as more than one respondent stated more than one area of expertise. The different areas of expertise were counted as separate entries.

Area of Expertise	Number (%)
Biomedical engineering	9/29 (31.0 %)
Clinical engineering	6/29 (20.7 %)
Medical devices & Instrumentation design	7/29 (24.1 %)
Life cycle management of MDs	4/29 (13.8 %)
Health technology assessment	2/29 (6.9 %)
Other	1/29 (3.5 %)

average years of expertise of the responders were 18.6 years and the standard deviation 12.2 years.

Consensus on Criteria

The median, interquartile range (IQR) (i.e., the range containing 50% of the data, obtained by subtracting the first quartile from the third quartile) and its interpretation, and the number of panelists for each recommendation are shown in Additional File 2. All the criteria reached the consensus threshold.

Reliability Analysis

Six out of seven domains showed a high internal consistency (see Table 2.4). This means that all the identified domains included a coherent and relevant set of criteria. Only the User Type domain showed a low ordinal alpha (0.58). In this case, we also performed further analysis to study how the ordinal alpha would change if one criterion would be omitted. As a result, the ordinal alpha reached an acceptable value of 0.76 by not considering the easiness of use.

Validity

The pilot study helped refine how questions were presented and one criterion was added during the piloting. Furthermore, although the respondents to the final survey suggested some additional criteria through the open questions, none of these was supported and shared by more than 3/18 respondents (16.7%, <30% cut-off).

Ranking and Correlations

All the calculated relative indexes were over 0.5, implying that all the criteria raised a great interest in the panelists. Also, all of the indexes were very close to each other,

TABLE 2.4

Summary of the reliability analysis.

Statistics	User Type	HTM	Design	Reliance on ext. factors	Material	Cost	Lifetime
Number of replies	24	25	25	20	25	25	25
Reliability of scale	0.58	0.92	0.7	0.9	0.96	0.9	0.84
Interpretation	Poor	Excellent	Acceptable	Excellent	Excellent	Excellent	Good
# of criteria	4	6	3	5	2	3	2
Changes applied after pilot survey	1 criterion added	-	-	-	-	-	-

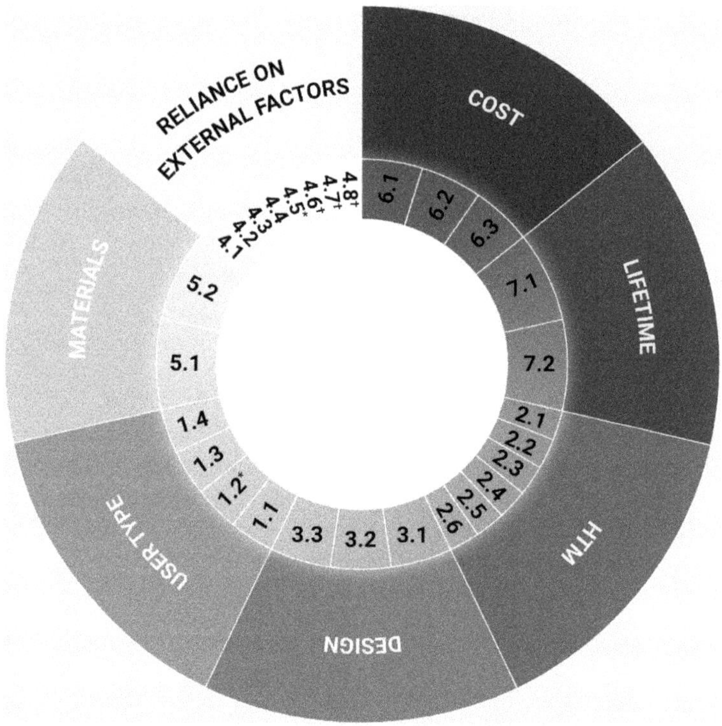

FIGURE 2.5
The final framework containing the domains and the criteria ranked according to the level of importance. The domains, although approximately sharing the same importance, are presented in descending order of importance starting from cost, clockwise. Each criterion is presented in the same descending order within each domain, too. *Criterion 1.2 is the criterion that was excluded after reliability/internal consistency analysis. *Criterion 4.5 is the criterion that was excluded because deemed of Medium importance. †Criteria 4.6, 4.7 and 4.8 are the criteria added *a posteriori* of our field studies and they are not currently ranked. The legend for the criteria in this figure can be found in Table 2.5.

highlighting the great importance of each domain, which in decreasing order were 0.888 for cost, 0.848 for lifetime, 0.84 for HTM, 0.832 for design, 0.824 user type, 0.824 for materials, and 0.792 for reliance on external factors (see Figure 2.5).

In particular, the next sections show the results grouped by domain (see Table 2.5 and Figure 2.6).

Cost: Maintenance cost (6.1), running costs (6.2), and initial cost (6.3) are "Fundamental", although the initial cost is seen as slightly less important with respect to the first two.

Lifetime: Both the lifetime of the MD (7.2) and its parts/components (7.1) are "Fundamental" and correlated. The lifetime of a MD depends on the functionality of its parts, in fact, more components are likely to cause more breakdowns.

TABLE 2.5

The correlations of the criteria grouped by domain. Correlation was reported only if "strong" (i.e., gamma greater than 0.5) and if the respective p-value was less than 0.05. The number specified in the correlation column refers to that of the correlated criteria. The "+" indicates a positive correlation, the "-" a negative correlation. †Criteria 4.6, 4.7, and 4.8 are the criteria added *a posteriori* of our field studies and they are not currently ranked. For this reason, correlations are also not calculated or reported (NA).

Criteria	Correlation
6. Cost	
6.1 Maintenance costs	6.2,6.3(+)
6.2 Running costs	6.1(+)
6.3 Initial cost	6.1(+)
7. Lifetime	
7.1 Lifetime of MD parts/components	7.2 (+)
7.2 MD lifetime	7.1(+)
2. HTM	
2.1 Need for consumables	2.2,2.3,2.4,2.5(+)
2.2 Need for spare parts	2.1,2.3,2.4,2.5(+)
2.3 Installation requirements	2.1,2.2,2.4,2.5(+)
2.4 Maintenance complexity	2.1,2.2,2.3,2.5,2.6(+)
2.5 Maintenance frequency	2.1,2.2,2.3,2.4(+)
2.6 Compatible consumables/spare parts	2.4(+)
3. Design	
3.1 Portability, compactness, robustness	3.2(+)
3.2 Limiting the number of components/spare parts	3.1(+)
3.3 Reusability	-
1. User type	
1.1 End users' background	1.4(+)
1.2 Easiness of use	-
1.3 Training needs	1.4(+)
1.4 User's understanding of the technical and clinical impact	1.1(+)
5. Material	
5.1 Durability of the material	5.2(+)
5.2 Robustness of the material	5.1 (+)
4. Reliance on external factors	
4.1 Reliance on power sources	4.2,4.3,4.4(+)
4.2 Reliance on water distribution	4.1,4.3,4.4(+)
4.3 Reliance on medical location air	4.1,4.2,4.4,4.5(+)
4.4 Need for sample preparation	4.1,4.2,4.3,4.5(+)
4.5 Understanding/stating the dependence of the MD from the medical location characteristics	4.3,4.4(+)
4.6 Resilience to dusty environments†	NA
4.7 Resilience to high-temperature environments†	
4.8 Resilience to high-humidity environments†	

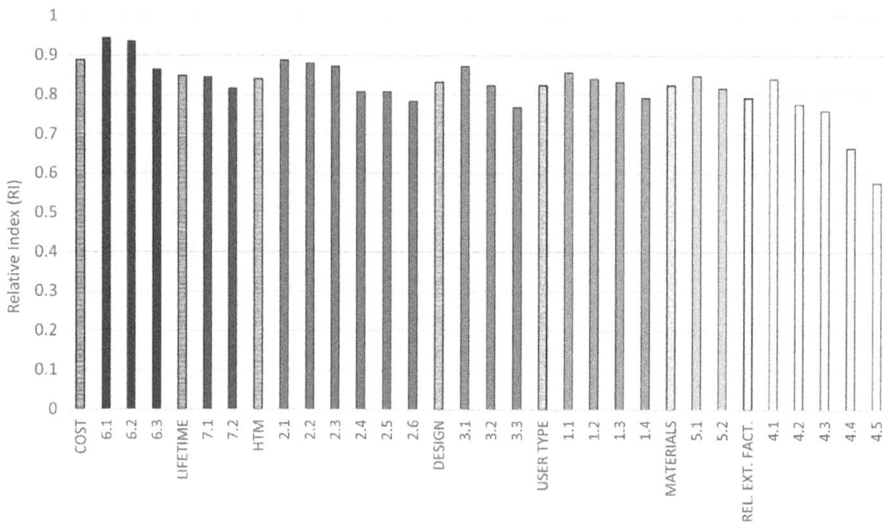

FIGURE 2.6
The relative indexes are reported for each criterion, grouped by domain. The first column of each group is the relative index of the domain itself. The full name of the criteria could not be reported for brevity. The legend is presented in Table 2.5.

HTM: The need for consumables (2.1) is number one in this domain and is positively correlated with the need for spare parts (2.2), the installation requirements (2.3), and the maintenance complexity (2.4) and frequency (2.5). This stresses recurring problems reported in the literature, in which the lack of spare parts, consumables, and functional maintenance are very common (8–10).

Design: Portability, compactness, robustness (3.1), and limiting the number of components/spare parts (3.2) are ranked as "Fundamental" and are correlated. Portability and compactness are greatly influenced by the number of components or spare parts. Limiting the latter would improve the durability of a functional device because fewer components are likely to reduce breakdowns.

User type: The end users' background (1.1) ranks at the top of this domain and is positively correlated with the users' understanding of the technical and clinical impacts (1.4). The latter is dependent on the former: a solid background is key to understanding the technical and clinical impacts. With regards to the easiness of use (1.2), which negatively correlated with the end user background (1.1) (although not significantly), it was discarded because of how it negatively affected the overall reliability of the user type domain. In this case, this variable could be re-assessed by the panelists to see if the results may change.

Materials: Both the durability (5.1) and the robustness (5.2) of the material are "Fundamental" and are correlated.

Reliance on external factors: Reliance on power sources (4.1) comes first in this domain, followed by the reliance on water distribution (4.2), the reliance on location air (4.3), and the need for sample preparation (4.4), with which it is correlated. All these are essential to the efficient and effective use of certain MDs. Regarding the understanding and stating the dependence of the MD from the medical location characteristics (4.5) indicated a low RI value of 0.576. Therefore, the Delphi study presented a consensus of the criterion being of "Medium" importance and the authors decided to discard it. Three additional criteria were added *a posteriori* of our field studies, i.e., resilience to dusty environments (4.6), resilience to high-temperature environments (4.7), and resilience to high-humidity environments (4.8).

Validation and Use of the Framework

The protocol (see Additional File 3) to test the framework includes an information table with the qualitative scoring of the MDs. The criteria selected only as informative and not discriminatory of a technology belong to the cost and lifetime domain. This further division was introduced because the comparison for such domains could only be made if there were references to use. For example, if someone were trying to compare two MDs with similar functions, they would need a MD of reference in order to consider the cost and life domains discriminatory. In this scenario, it would then be possible to compare the costs and the lifetime of one of the MDs compared to the one of reference and consequently assign those values such as "low", "medium", or "high".

The three criteria added *a posteriori* are "Resilience to dusty environments", "Resilience to high-temperature environments", and "Resilience to high-humidity environments".

According to the traffic light rating system adopted, poor performances were marked as red, medium performance as yellow, and good performance as green.

For example, as regards the "End users' background", red would denote the fact that the technology could only be used by medical doctors, yellow by any healthcare professionals, and green by lay users. Or as regards the "Reliance on power sources", red would denote the fact that the technology relies only on the power supply, yellow on the power supply and batteries, and green on the latter and solar panels. Applying this process to one of the selected MDs, i.e., the pulse oximeter, to make further examples, we can notice how it scores yellow for "End users' background", as lay users are not enlisted among possible end users. Such a device also scores green for "Training needs", as no further training is required, and scores red (also the missing data are classed as red) for resilience to dusty, high-temperature and -humidity environments due to the manufacturer not providing any information about these problematic environments.

The selected MDs were: low-cost computed tomography scanner, pulse oximeter, anesthesia delivery for LRSs, mobile-enabled non-invasive measure-through motion and low perfusion pulse oximeter, electrocardiogram handheld digital, phototherapy for jaundice, external fixation for bone fracture, and white blood cell counting system.

Only nine out of the 21 discriminatory criteria performed generally well (predominantly green cells) and they belonged to different domains: installation requirements, maintenance frequency and complexity, portability, compactness and robustness, limiting the number of components/spare parts, reusability, reliance on water distribution, reliance on medical location air, and need for sample preparation.

The three MDs that resulted in efficient performance are the two pulse oximeters and the external fixation for bone fracture. The four MDs, whose efficiency resulted medium performance, are the electrocardiogram, the phototherapy device and the white blood cell counting system. Finally, the computed tomography scanner and the anesthesia machine resulted in not very efficient performance.

Discussion

This study aimed to respond to the urgent need for a comprehensive framework for the design of MDs resilient to low-resource settings (LRSs) and to propose a possible solution. To date (April 2021), MD design frameworks have been proven ineffective when using the MDs outside of "their bubble", i.e. settings that have more resources and some guaranteed standards that are not readily available in LRSs (38). The results of this study show the timeliness and importance of such a topic. In fact, the international experts' views were elicited through MCDA, which was proven to aid decision-making processes, knowledge production and consumption (39), being able to capture all the various dimensions of the evaluation problem (40), enhancing the quality of decisions, allowing them to be more rational and efficient and transparent (39, 41, 42), also in the relevant emerging technology and biomedical fields (43).

From the results, it emerged that the experts in different fields shared their opinions and agreed on most of the criteria. A particular criterion had a slightly lower response rate of 20/25, with five panelists who replied, "Do not know". The authors think that this criterion, i.e., Understanding/stating the dependence of the MD from the medical location characteristics (i.e., Group 0,1,2), was underestimated probably because of a general non-expertise in such a specific topic.

Correlations are essential to understand how criteria are interrelated. In fact, they give an understanding of the trends that characterize the set of criteria. Positive correlation between criteria A and B means that the higher the importance of criterion A is also followed by higher importance for criterion B. The opposite is true in the case of negative correlation. This type of information is useful to inform the design of the MDs, since the criteria with positive correlations need to be jointly taken into account to guarantee the successful design and development of a MD. For example, in the domain of cost, running costs are positively correlated with maintenance costs, but not to the initial costs. This confirms that the respondents consider these two kinds of costs interrelated. Hence, they need to be jointly considered when designing a MD.

From the results, it is clear that even if all the domains share a similar relative index of importance, the resulting priorities are a comprehensive representation of the current state of the art (as of 2021) regarding the design of MDs for LRSs. In particular, the domain of cost comes first, and the domain of the reliance on external factors comes last. However, within the domain of cost, the operating and maintenance costs resulted in being slightly prioritized in respect to the initial cost. Although capital costs can be high for some equipment, the operating or maintenance costs are often underestimated. Malkin (12) had concluded that for a MD to be effective and economical at point of purchase is not sufficient for a technology to help LRSs, highlighting common misconceptions such as the belief that the capital cost of a technology is always the primary barrier.

Indeed, the cost is crucial, but the underestimation of the reliance on external factors or of the user type (i.e., the last but two) is dangerous. These factors are the ones that may severely influence the functional operation of MDs, as it has also been reported in the literature, and that leads to the so-called "contextualized design".

For this reason, as mentioned before, after their latest field studies in Benin and Uganda, the authors of this paper decided to include three new criteria that underline critical challenges for LRSs, i.e., the presence of dusty environments, high temperatures, and high humidity. Once again, this further highlights the importance of these factors that are often overlooked.

In the authors' opinion, the results from this study are a relevant representation of the current (2021) approach of MD designers. Clearly, there is a need to change the way of approaching MD design if we want to support global health without inequalities across the countries.

This study is the first of its kind to use a triangulation of qualitative and quantitative methods to obtain a comprehensive and effective framework for the design of MDs resilient to LRSs. This framework is further proof of the need for a change and adaptation both of the existing international standards and the MD design criteria. The authors believe that the approach presented by Adler (18) and currently utilized by the IT industry, i.e., the use of the IP code standard and military standards (MIL-STD-810H) is the future of MD design. The harsh conditions typical of LRSs, in fact, could be simulated through apposite tests for humidity, temperature, pressure, and others adapted from the military world.

Furthermore, any researcher or designer involved in MD design could potentially use this framework either to check if an existing design is compliant or to start a new innovative and/or conceptual design.

The validation of this framework proved that the protocol to follow for assessing a MD is quite simple and does not require external expert inputs. Any person with a biomedical engineering background should be able to complete the framework and straightforwardly assess the technology. The validation also confirmed that the domain of reliance on external factors is often underestimated and resulted in obtaining the majority of yellow and red scores.

Limitations of the Study

As already mentioned, some of the criteria were classified as informative, because, in order to use such domains for comparisons, there is a need for extra references, which can be investigated in case someone wanted to compare different options of the same kind of technology.

In addition, some of the scales are only available as qualitative, as semi-quantitative scales would need to be investigated further with *ad-hoc* studies, and there are several "-", meaning that no information is available to score some MDs. The latter are data gaps that should be filled to get a more realistic understanding of the performances. At the time of writing, missing data/gaps were assigned the worst-case color (i.e., red), using a precautionary approach.

Finally, the validation carried out is a valid internal validation of the framework. However, an additional external validation/testing could preferably be performed as well.

Conclusions

The aim of this study was to highlight current (2021) challenges and gaps in the design process of MDs, to highlight the urgency in addressing such gaps, and to create a framework for contextual and user-driven design of MDs. Such a framework will facilitate design and production of MDs which will perform correctly and efficiently in any country in the world, irrespective of the available income and/or resources. Although frameworks exist, such as the frugal framework of Aranda et al. (13), the focus here is the aspects of MDs on which the designer has a potential influence.

The advantages of the framework presented in this paper are the fact that it emphasizes the urgency of taking into consideration particular realities and the technical tool it offers. Such a tool, in fact, can be one of the first steps towards this inclusive approach. Nonetheless, it is pivotal to underline that the authors' invitation to take contextual realities into account should not be misinterpreted as "relativism", because trying to defend and safeguard differences will lead to failing to see any common prospect.

Rather, this framework fosters glocalisation, as per Bauman's definition (44): i.e., the preservation of individual identities within a complex system. In fact, far from dividing the world into different parts, each characterized by the richness of its peculiarities (none being superior to the others), we could reaffirm Augé's theory of the "Non-place", i.e., an effectively inclusive space where each context finds its citizenship (45).

To conclude, this framework could encourage and inform the creation of novel, contextualized international standards, which would empower LMICs by enforcing import restrictions according to the WTO-TRIPS, by fostering the local production of medical equipment and by obliging MD manufacturers to comply with this new consensus standard if they wanted to market their products in LMICs.

Starting from this paper, further international political actions should follow to bring forward design frameworks based on the best available evidence and real users' needs elicited by experts, in order to drive the real change of the existing norms, regulations, and standards.

Additional Files

Additional file 1 is a copy of the final version of the questionnaire (.pdf).

Additional file 2 is a Word file (.docx) and is a table containing the design criteria with median, interquartile range, its interpretation, and the number of panelists.

Additional file 3 is an Excel file (.xlsx) and is the tool that was used for validating the framework and that can be used for assessing other technologies.

Declarations

Ethics Approval and Consent to Participate

We obtained clearance to conduct this study from the University of Warwick Ethical BSREC Committee (REGO-2018-2283). Additionally, we obtained consent from every participant prior to data collection.

Acknowledgements

The authors would like to acknowledge Luis Montesinos (Assistant Professor at the Tecnológico de Monterrey, Mexico) for the translation of the questionnaire in French. The authors would like to acknowledge Hardip Boparai (PhD student at the University of Warwick, United Kingdom) and Katy Stokes (PhD student at the University of Warwick, United Kingdom) for proofreading this manuscript and for their valuable suggestions.

Note

1 D. Piaggio, R. Castaldo, M. Cinelli, S. Cinelli, A. Maccaro, L. Pecchia, *A MCDA framework for designing medical devices resilient to low-resource settings: an application for*

low- and middle-income countries (LMICs), in «Globalization and Health» 17, 64 (2021) https://globalizationandhealth.biomedcentral.com/articles/10.1186/s12992-021-00718-z

References

(1) WHO. World Bank and WHO: Half the world lacks access to essential health services, 100 million still pushed into extreme poverty because of health expenses. 2017. Available from: .
(2) Arasaratnam A, Humphreys G. Emerging economies drive frugal innovation. *World Health Organization Bulletin of the World Health Organization*. 2013;91(1):6.
(3) Zhang W, Chatwin C, Lui R. The Chinese medical device market: Market drivers and investment prospects. *Journal of Commercial Biotechnology*. 2016;22(2):33–39.
(4) WHO. *Barriers to innovation in the field of medical devices*: WHO; 2010.
(5) Gauthier AK, Cruz G, Medina L, Duke S, editors. Design factors for medical device functionality in developing countries. In *IIE Annual Conference Proceedings*: Institute of Industrial and Systems Engineers (IISE); 2013 (p. 2227). Institute of Industrial and Systems Engineers (IISE).
(6) Piaggio D, Medenou D, Houessouvo RC, Pecchia L. Donation of medical devices in low-income countries: Preliminary results from field studies. In Badnjevic A, Škrbić R, Gurbeta Pokvić L, editors. *International Conference on Medical and Biological Engineering*: Springer; 2019, May, pp. 423–427.
(7) Di Pietro L, Piaggio D, Oronti I, Maccaro A, Houessouvo RC, Medenou D, et al. A framework for assessing healthcare facilities in low-resource settings: Field studies in Benin and Uganda. *Journal of Medical and Biological Engineering*. 2020 Aug;40:526–34.
(8) Shah MT, Joshipura M, Singleton J, LaBarre P, Desai H, Sharma E, et al. Assessment of the availability of technology for trauma care in India. *World Journal of Surgery*. 2015;39(2):363–372.
(9) LaVigne AW, Triedman SA, Randall TC, Trimble EL, Viswanathan AN. Cervical cancer in low and middle income countries: Addressing barriers to radiotherapy delivery. *Gynecologic Oncology Reports*. 2017;22:16–20.
(10) Taylor A, Muelenaer P, Kochersberger K, Muelenaer A, Bickford L, Redican K. Innovating for global health: Study of healthcare technology failure in southern Malawi. *Annals of Global Health*. 2016;82(3):600–601.
(11) Rusatira JC, Tomaszewski B, Dusabejambo V, Ndayiragije V, Gonsalves S, Sawant A, et al. Enabling access to medical and health education in Rwanda using mobile technology: Needs assessment for the development of mobile medical educator apps. *JMIR Medical Education*. 2016;2(1).
(12) Malkin RA. Barriers for medical devices for the developing world. *Expert Review of Medical Devices*. 2007;4(6):759–763.
(13) Aranda JC, Jagtap S, Moultrie J. Towards a framework for holistic contextual design for low-resource settings. 2016.
(14) Richards-Kortum R, Oden M. Devices for low-resource health care. *Science*. 2013;342(6162):1055–1057.

(15) Pecchia L. *Health technology assessment of medical devices in low and middle income countries: Study design and preliminary results*. EMBEC & NBC 2017: Springer; 2017. pp. 225–228.
(16) Nimunkar AJ, Baran J, Van Sickle D, Pagidimarry NK, Webster JG, editors. Medical devices for developing countries: Design constraints and approaches. In 2009 Annual International Conference of the IEEE Engineering in Medicine and Biology Society 2009 Sep 3 (pp. 7048-7051). IEEE.
(17) Zeschky M, Widenmayer B, Gassmann OJR-TM. Frugal innovation in emerging markets. Research-Technology Management. 2011 Jul 1;54(4):38–45.
(18) Adler D, Heller O, Vosk M. Medical devices for military use. Military medicine. 1993 Oct 1;158(10):677–680.
(19) WHO. *WHO compendium of innovative health technologies for low resource settings, 2011–2014: Assistive devices, eHealth solutions, medical devices, other technologies, technologies for outbreaks*: World Health Organization; 2015.
(20) Pope C, Mays N. Qualitative research: Reaching the parts other methods cannot reach: An introduction to qualitative methods in health and health services research. BMJ. 1995 Jul 1;311(6996):42–45.
(21) Goddard W, Melville S. *Research methodology: An introduction*: Juta and Company Ltd; 2004.
(22) Carter N, Bryant-Lukosius D, DiCenso A, Blythe J, Neville AJ, editors. The use of triangulation in qualitative research. In *Oncology Nursing Forum*: Oncology Nursing Society. 2014 Sep (Vol. 41, No. 5, pp. 545–7).
(23) Ritchie J, Lewis J, Nicholls CM, Ormston R. *Qualitative research practice: A guide for social science students and researchers*: SAGE; 2013.
(24) Etikan I, Musa SA, Alkassim RS. Comparison of convenience sampling and purposive sampling. American journal of theoretical and applied statistics. 2016 Jan 5;5(1):1–4.
(25) Polisena J, Castaldo R, Ciani O, Federici C, Borsci S, Ritrovato M, et al. Health technology assessment methods guidelines for medical devices: How can we address the gaps? The International Federation of Medical and Biological Engineering perspective. *International Journal of Technology Assessment in Health Care*. 2018;34(3):276–289.
(26) IFMBE. Kick-off of the First African Scientific Society of Medical and Biological Engineering. 2018. Available from: http://htad.ifmbe.org/kick-off-of-the-first-african-scientific-society-of-medical-and-biological-engineering/
(27) Okoli C, Pawlowski SD. The Delphi method as a research tool: An example, design considerations and applications. *Information & Management*. 2004;42(1):15–29.
(28) Investopedia. Bandwagon Effect. Available from: www.investopedia.com/terms/b/bandwagon-effect.asp
(29) Wikipedia. Bandwagon Effect. Available from: https://en.wikipedia.org/wiki/Bandwagon_effect
(30) Revelle W, Condon DM. Reliability from α to ω: A tutorial. Psychological assessment. 2019 Dec;31(12):1395.
(31) Cinelli M, Coles SR, Sadik O, Karn B, Kirwan K. *A framework of criteria for the sustainability assessment of nanoproducts*. Journal of Cleaner Production. 2016 Jul 10;126:277–287.
(32) Furr M. *Scale construction and psychometrics for social and personality psychology*: SAGE; 2011.

(33) Bonanomi A, Ruscone MN, Osmetti SA, editors. The Polychoric Ordinal Alpha, measuring the reliability of a set of polytomous ordinal items. In *SIS 2013 Conference: Advances in latent variables: Methods, models and applications*, Brescia, Italy; 2013.
(34) Zumbo BD, Gadermann AM, Zeisser C. Ordinal versions of coefficients alpha and theta for Likert rating scales. *Journal of Modern Applied Statistical Methods*. 2007;6(1):4.
(35) Wee I, Varughese RS, Syn N, Choong A. Non-operative management of Type A Acute Aortic Syndromes: A systematic review and meta-analysis. *European Journal Vascular and Endovascular Surgery*. 2019;58(1):41–51.
(36) WHO. *WHO compendium of innovative health technologies for low-resource settings: 2016–2017: Medical devices, eHealth/mHealth, medical simulation devices, personal protective equipment, assistive products, other technologies*: World Health Organization; 2018.
(37) Pugh S, editor. Concept selection: A method that works. In *Proceedings of the International conference on Engineering Design*; 1981 Mar 9 (pp. 497–506).
(38) Pecchia L, Piaggio D, Maccaro A, Formisano C, Iadanza E. The inadequacy of regulatory frameworks in time of crisis and in low-resource settings: Personal protective equipment and COVID-19. *Health and Technology*. 2020 Nov;10(6):1375–83.
(39) Carayannis EG, Ferreira JJ, Jalali MS, Ferreira FA. MCDA in knowledge-based economies: Methodological developments and real world applications. *Technological Forecasting and Social Change*. 2018;131:1–3.
(40) Carayannis EG, Goletsis Y, Grigoroudis E. Composite innovation metrics: MCDA and the Quadruple Innovation Helix framework. *Technological Forecasting and Social Change*. 2018;131:4–17.
(41) González A, Ortigoza E, Llamosas C, Blanco G, Amarilla R. Multi-criteria analysis of economic complexity transition in emerging economies: The case of Paraguay. *Socio-Economic Planning Sciences*. 2019;68:100617.
(42) Lerche N, Wilkens I, Schmehl M, Eigner-Thiel S, Geldermann J. Using methods of multi-criteria decision making to provide decision support concerning local bioenergy projects. *Socio-Economic Planning Sciences*. 2019;68:100594.
(43) Chen S-H, Lin W-T. Analyzing determinants for promoting emerging technology through intermediaries by using a DANP-based MCDA framework. *Technological Forecasting and Social Change*. 2018;131:94–110.
(44) Bauman ZJTE. On glocalization: Or globalization for some, localization for some others. 1998;54(1):37–49.
(45) Augé M. *Non-lieux. Introduction à une anthropologie de la: Introduction à une anthropologie de la surmodernité*: Le Seuil; 2015.

A Framework for Assessing Healthcare Facilities in Low-Resource Settings: Field Studies in Benin and Uganda[1]

Introduction

United Nations' (UN) 2030 Agenda for Sustainable Development includes 17 Sustainable Development Goals (SDGs) which are an urgent global call for action (1). Sustainable Development Goal 3, Good health and well-being, strives for ensuring healthy lives globally, emphasizing the access to quality essential healthcare services, which is currently inequitable at the expenses of low- and middle-income countries (LMICs) (1, 2).

During the Sixtieth World Health Assembly, the World Health Organization (WHO) recognized the severe implications of the inappropriate provision of health technologies to support the quality of health in developing countries. The lack of appropriate technologies was recognized as a barrier to the achievement of the SDGs, specifically SDG3 with its specific target of Universal Health Coverage (UHC), which impacts on the guarantee of human rights (3).

Patients and healthcare providers in LMICs have limited access to medical devices (MDs) (4). One barrier for the accessibility to medical technologies is the lack of funds and the high costs of imported MDs. However, if this had been the main and only problem, donations would have been a solution (5). In this regard, the WHO estimates that more than 80% of medical equipment in LMICs is donated (6), but only 10-30% of this is operating (7) due to the lack of spare parts (8), of specialized personnel (9, 10), of a good maintenance and management systems (5, 11), and due to the harsh environmental conditions.

Other more impelling barriers derive from the fact that medical equipment is largely designed for and manufactured in high-income countries (HIC), following good manufacturing practices, international standards and minimum requirements, optimized for well-structured facilities, rendering it difficult to be implemented in low-resource settings (LRSs) (12). Medical equipment is designed to operate in a clean, sterile, climate-controlled environment, with a reliable main power supply, a good maintenance system, a working supply chain, and expert healthcare operators.

DOI: 10.1201/9781003436348-9

However, in LMICs, the safe and efficient use of MDs is hindered by harsh environmental and climatic conditions (e.g., high temperatures, humidity, and dust), unreliable main power supplies and poor infrastructures (13, 14). These conditions cause frequent failures and trigger a higher demand for spare parts, which are expensive and difficult to find, making the maintenance of MDs as problematic as their acquisition (15). As regards to local hospitals in LMICs, the challenges are exacerbated by the underlying structures of the buildings (16–21) (see Online material 1–3).

Given the paucity of information on the conditions of LRSs, several authors have performed reviews, interviews, and observations aimed at assessing the adequateness and appropriateness of local hospitals to deliver surgical care (16, 17), or intensive care (18–21), or to manage non-communicable diseases (NCDs) (22) such as hypertension (23). In 2011, Hsia et al. (16) ran a survey based on the WHO list of essential surgical services, analyzing hospitals in Kenya, Rwanda, Tanzania, Uganda, and Ghana, focusing on different variables, including basic infrastructure, medicine storage capability, and quality systems. The authors concluded that none of the analyzed countries had the proper infrastructures for delivering surgical care. The same problems along with supply chain difficulties in terms of equipment and supplies, old infrastructures not big enough for the ever-growing population, a limited number of beds and monitored beds, were identified by Albutt et al. (17). They also stressed the fact that the equipment, if present, is outdated and works intermittently, lacking maintenance. In some cases, the situation is exacerbated by the differences between rural and urban areas (21, 22).

Accordingly, the authors of this paper decided to implement and test a systematic framework, consisting of semi-structured interviews with qualified personnel and direct measurements campaigns, for the assessment of health centers and hospitals in LRSs, given the lack of a standard protocol and of tangible, quantifiable, and comparable information regarding these settings.

This paper presents the framework and its application during two field studies performed in Uganda and Benin.

Two Countries of the Developing World: Uganda and Benin

The UN created the Human Development Index (HDI) to evaluate the development of a country, based on different factors, including economic growth, life expectancy at birth, education, and the standards of living (24). A developing country, or a LMIC, is a country with few resources, a low HDI compared to other countries, and a gross national income per capita (GNI) below 4035$.

Uganda, an Eastern African country with a GNI of 620$, and Benin, a Western African country with GNI of 870$, are both ranked as low-income countries (LICs) and are among the 25 poorest countries in the world (25). Table 2.6 summarizes some relevant data for both countries in comparison with a typical HIC.

TABLE 2.6

Relevant data for Uganda, Benin, and Italy (26–29).

	Uganda	Benin	Italy
Area	241,037 km^2	114,763 km^2	301,338 km^2
HDI	0.528	0.520	0.883
Population	42.86 million	11.49 million	60.48 million
GNI	620$	870$	34,456$
HIV[a] prevalence	5.7% (15–49 years old)	1.13% (15–49 years old)	0.3% (15–49 years old)
Life expectancy at birth	60 years	62 years	83 years
Percentage of the population using an improved drinking water source	79%	77.9% (rural and urban)	100%
Physicians density (physicians per 1000 inhabitants)	0.09	0.16	3.95
Houses with centralised electricity supply	22%	43%	100%
Quality of electricity	3.43/7	2.06/7	5.91/7
Mobile phone subscriptions (per 100 people)	57.27	82.38	137

[a] Human immunodeficiency virus

The Ugandan National Health System

Uganda is organized into four administrative regions (i.e., Northern, Eastern, Central, and Western), which are further divided into 134 districts and one city (the capital city of Kampala). The National Health System (NHS) in Uganda comprises both a private and a public sector. The public sector includes all Government of Uganda health facilities under the Ministry of Health (MoH) and other ministries. The private health sector, which plays an important role in the delivery of health services in Uganda, includes Private Not for Profit, Private Health Practitioners, and Traditional Contemporary Medicine Practitioners. The provision of the health system is decentralized with districts and health sub-districts playing a key role in the delivery and management of health services at those levels. The health services are structured into National Referral Hospitals (NRHs) and Regional Referral Hospitals (RRHs), Fourth level General Hospital Health Centers (HCs), Third level HCs, Second level HCs, and Village Health Teams (30).

The Beninese National Health System

Benin is divided into 12 departments, 77 communes, and 546 districts, all referring to 34 sanitary zones. Its most important departments are Oumé, with the official capital

(Porto-Novo), and Littoral, where the economic and administrative capital (Cotonou) is located.

Specifically, also for Benin, the NHS is decentralized, based on a pyramidal structure comprising of three levels:

1. **Central:** the MoH and General Secretariat are in charge of defining policies, strategies, and directives;
2. **Intermediate:** the Departmental health directorates are in charge of implementing and coordinating the governmental health policies;
3. **Peripheral:** including Health zones, Commune Health Centers, village health units and private hospitals, which are the operational units.

The multifaceted realities of the departments are difficult to map out and present differences, mainly between the north and the south of the country. In addition to the ordinary health structures, there are numerous vocational hospitals spread around the country (31).

Methods

Selection of health facilities.

Benin and Uganda were selected as first destinations, because of our previous experiences and the established networks. Local contacts were able to link us with the hospitals, where we performed our assessment.
The selected health structures were:

Kawolo General Hospital (Buikwe, Uganda) (H1)
Mengo Hospital (Kampala, Uganda) (H2)
Naguru General Hospital (Kampala, Uganda) (H3)
Hôpital La Croix (Zinviè, Benin) (H4)
Centre Hospitalier Universitaire (CHU) de Zone Suru-Léré (Cotonou, Benin) (H5)
CHU de Zone d'Abomey-Calavi (Abomey-Calavi, Benin) (H6)

Questionnaire Preparation and Validation

The questionnaire was drafted during focus groups among experts of medical device design, contextualized design, health technology assessment and management, and clinical engineering (see Figure 2.7). During these focus groups, relevant literature was used and integrated by the experts' knowledge to define the different

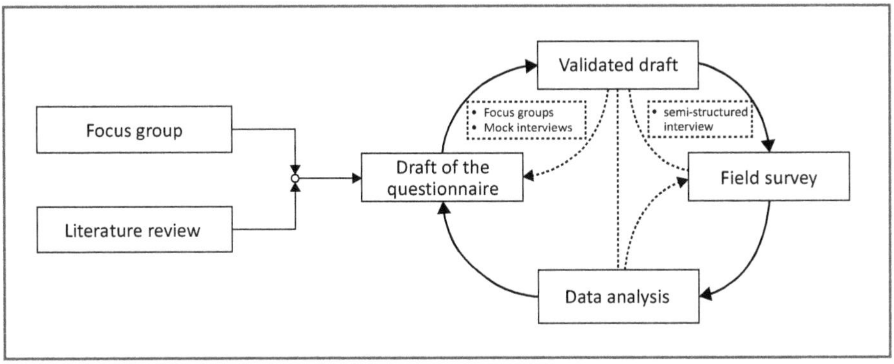

FIGURE 2.7
The diagram shows the methodologies used for the development and administration of the questionnaire.

questions and sections of the questionnaire: in particular, the sections on the general characteristics of the facility, electrical access, human resources, and facility environment were based on refs. (32, 33), those on medical electrical equipment on refs. (34, 35), and the section on patient data management on ref. (31).

At first, the questionnaire was conceived to be online, but we realized that more relevant information could be captured if it was administered in person as a semi-structured interview (see Online material 4). The first draft of the questionnaire was circulated among the participants of the first focus group for internal validation and possible corrections. Once the final version was ready, it was tested for further validation on an anesthesiologist with working experience in hospitals in Sierra Leone (with Emergency).

Electrical Safety Measures

Electrical safety measures were carried out in accordance with IEC 62353-2015, using an electrical safety analyzer (ESA620) by Fluke (see Online material 5). The MDs analyzed were selected depending on availability at the health centers at the time of the assessment. Our inspection protocol included:

- Voltage and earth check on the main power supply;
- Visual inspections to check the integrity of the devices, the cables, and the accessories;
- Protective earth measurement;
- Insulation resistance measurement;
- Leakage currents measurement;
- Functionality tests (e.g., ECG wave simulation).

Results

Questionnaire

The final questionnaire (see Online Material 6) consisted of 71 questions, organized in nine sections: Introduction and Authorization, Personal Information, Facility Information, Facility General Characteristics, Electrical Access, Human Resources, Facility Environment, Medical Electrical Equipment, and Database. The questions were of different types (e.g., Yes/No, multiple choice, etc.) and some of them were quantitatively assessing some dimensions, and others were assessing how some dimensions were perceived by the interviewee. The latter had six possible answers, based on a five-step Likert-type scale, i.e., "Very low", "Low", "Middle", "High", "Very high", "Do not know".

The six interviewees, one per hospital, were all males (21-50 years old), had a college or university degree, and comprised four biomedical engineers, one medical doctor, and one nurse. All had been working in those roles for an average of 7 years.

General Information About the Facility

All the assessed hospitals are public and third-level structures (urban), except the one in Zinviè, which is private and a second-level structure (semi-urban). All the hospitals, but H1, rely on piped water. H1 relies on tanker water (also available in H2, H3, and H6), or water from a well (also available in H4 and H6). H1 also has structures for collecting rainwater. Only some of the structures have a functioning landline telephone (H2, H3, and H5). Other basic facilities are: a mobile phone (H1, H2, H3, and H5), a short wave radio (H2), a computer (all but H4), an internet service (H2, H3, H5, and H6), and an ambulance (all but H5). As regards the access to electricity, all the hospitals can rely on the central supply, most of them have a generator (H1, H2, H4, H5, and H6), and some of them have a solar electric system (H1, H2, and H6) (see Table 2.7 for further information). All the hospitals reported a certain degree of incompatibility among the local sockets and the plugs of donated MDs (see Figure 2.8).

Human Resources

All the facilities reported a chronic lack of doctors (one for more than 30 patients), clinical officers (one for more than 30 patients), of nurses (with an average of one nurse per 20 patients), and laboratory technicians. In general, most of the interviewees judged this situation as "poor" or "very poor".

Facility Environment

In general, most of the facilities were judged by the interviewee as poorly insulated or distant from dust, smoke, and undue noise. A better situation was found regarding the

TABLE 2.7

Summary of the information and the ratings of the electrical access, reliability, and safety.

	Hospital Power outages per month	Rating of the access to the main source of electricity	Rating of the quality and reliability of the electricity of the facility	Available and functional systems for electrical safety	Rating of the electrical safety in the facility	Rating of the compatibility of the working voltage and frequency required for the MDs and those available at the facility
H1	4–6	Acceptable	Poor	EG	Poor	Good
H2	1–3	Acceptable	Very good	EG, EN, IT	Good	Very good
H3	1–3	Good	Acceptable	EG, EN, IT	Very good	Good
H4	10+	Poor	Poor	EG, EN	Acceptable	Poor
H5	10+	Good	Poor	EG, EN, IT	Acceptable	Very good
H6	10+	Acceptable	Acceptable	EG, EN	Good	Very good

EG electrical grounding, *EN* equipotential node, *IT* isolation transformer

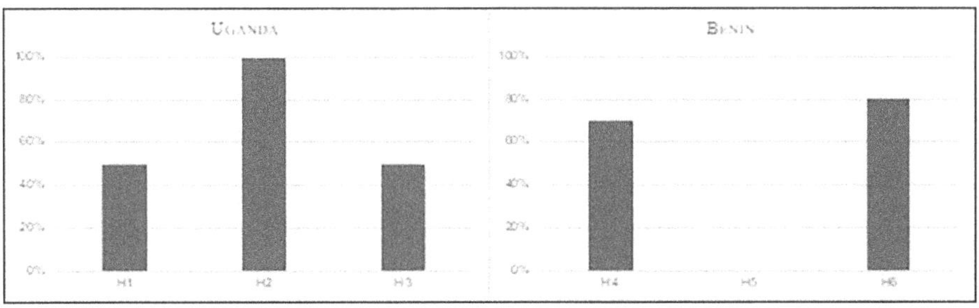

FIGURE 2.8

The compatibility between the local sockets and the plugs of donated medical devices for Uganda and Benin.

insulation or distance from foul odors, the lighting, and the adequateness of ventilation (see Table 2.8). Similarly, the disposal of different kinds of waste (e.g., non-risk waste, infectious waste, sharps, etc.) was mostly rated as "very good", apart from H1, for which it was rated as "very poor". All the hospitals' environments are cleaned five or more times per week.

Medical Electrical Equipment, Maintenance, and Management

Figure 2.9 shows the distribution of essential MDs and services within the six hospitals. The list is ordered from the least available piece of equipment to the most available one. Complex devices like colonoscopes, mammographs, CT-scanners and

TABLE 2.8

The ratings of the insulation or distance of the structures from undue noise, dust, foul odors, and smoke.

Hospital	Undue noise	Smoke	Dust	Foul odors
H1	Very poor	Very poor	Very poor	Very poor
H2	Very poor	Very poor	Very poor	Acceptable
H3	Very poor	Very poor	Very poor	Very poor
H4	Acceptable	Acceptable	Poor	Acceptable
H5	Good	Good	Acceptable	Good
H6	Good	Poor	Acceptable	Good

Country Hospital	Uganda H1	Uganda H2	Uganda H3	Benin H4	Benin H5	Benin H6
Colonoscope	0	1	0	0	0	0
Mammograph	0	0	1	0	0	0
CT-scanner	0	1	1	0	0	0
Gastroscope	0	1	2	0	0	0
Infant reanimation centre	0	0	1	0	1	1
X-Ray Machine	1	3	1	1	1	1
Ambulance	0	3	1	2	0	2
Defibrillator	2	4	NA	3	2	0
Ventilator ICU	0	4	8	0	2	1
Hemocytometer	1	4	1	8	4	1
Ultrasound machine	0	8	2	4	2	3
Oxygen systems/cylinders	10	10	1	0	0	0
Syring pump	0	8	0	2	8	8
Autoclave for sterilisation	2	8	10	2	4	1
Operating theatre with basic equipment	3	8	3	8	3	2
Suction pump	1	8	10	0	8	1
Infant warmer	0	8	8	4	8	0
Anasthetic machine	1	10	8	0	2	8
Fetal monitor	1	8	10	0	8	2
Neonatal incubator	8	10	1	3	4	4
ECG machine	0	4	8	10	8	2
Patient monitor	1	10	10	10	8	10
Scale for adult	0	10	10	10	10	10
Scale for newborns	0	10	10	10	10	10
Puls oximeter	2	10	10	10	10	10
Thermometer	8	10	10	10	10	10
Blood pressure machine/cuff	10	10	10	10	10	10

FIGURE 2.9

The distribution of essential MDs and services within the six hospitals. The ranges were substituted with the average value. The value 10 stands for 10 or more. Red circles individuate a low availability of the MD, yellow circles a medium availability, and green circles a high availability.

X-Ray machines are a rare find, compared to blood pressure machines, thermometers, pulse oximeters, scales, and patient monitors.

All the hospitals have a biomedical engineering department, except H1 and H4. The approach to MD maintenance varies depending on the facility. Most of the structures that have a biomedical engineering department in charge of the preventive and the corrective maintenance (H2, H5, and H6), with the exception of H3 that only follows corrective maintenance practices. However, the structures that do not have a biomedical engineering department (H1 and H4) are only relying on on-call biomedical engineers/technicians, thus following only corrective maintenance practices.

The most recurrent problematics were found to be the lack of funding, of essential MDs, of spare parts and consumables, of expertise. One of the hospitals (H1) also denounced the inexistence of a policy regulating the donation of MDs.

As regards the management, all the hospitals but one (H4) have a form of inventory: three hospitals have a paper-based one (H1, H2, and H3) and two have or are transitioning to a computerized one (H5 and H6). The most recorded information pertained to the year of manufacture, the type, the serial number, the year of acquisition, the class function, the date of service, the routine servicing, the reason for acquisition, and the technical characteristics.

Electric Safety Measurements

We were able to perform these measurements in four hospitals (one in Uganda and three in Benin). Regarding the intensive care unit of H2, the inspected sockets were up to standard (voltage on the mains of 238 V; voltage between the neutral and the ground of 0.2 V). In H2, out of the five tested devices (i.e., one defibrillator, three patient monitors, and one ECG), two of the patient monitors[1],[2] did not pass visual inspection because they were lacking respectively the blood pressure cuff, the ECG cables, and the power cable. Moreover, the ECG cables of the defibrillator[3] were not working, in fact, when inputting a simulated signal nothing was showing on the screen. All the other devices passed the inspection. Table 2.9 reports the findings on electrical safety.

As regards H4, the Hôpital La Croix in Benin, the sockets of the surgical room we inspected were up to standard (voltage on the mains of 240 V; voltage between the neutral and the ground of 2.6V).

H6 had functional sockets in the ambulatory, however the maintenance lab and two surgical theatres had no ground, with voltages between the neutral and the ground of 89.9-134.5 V. Throughout the different buildings we noticed that the grounding system was a common problem, along with the inversed polarity of the sockets. In H5, all but one of the inspected departments (i.e., Biochemistry, Hematology, HIV lab, and new equipment room) presented troubles with the earth system (with voltages between the neutral and the ground between 8 and 43 V), with the exception being the department of radiology. The report can be seen in Table 2.10.

TABLE 2.9

The report of the electric safety measurements done in Uganda's Mengo Hospital Intensive Care Unit (H2).

Equipment	Protective earth (ohm)	Insulation					Leakage current	
		Mains-PE	AP-PE	Mains-AP	Mains-NE	AP-NE	Equipment	AP
Defibrillator[3a]	NC	Infinite	NA	NA	NA	NA	0.2[b]	5.4[b]
Patient monitor[2]	0.075	Infinite	NA	NA	Infinite	NA	NA	NA
Patient monitor[1a]	–	–	–	–	–	–	–	–
Patient monitor	0.065	Infinite	Infinite	Infinite	Infinite	–	0.3	5.2
ECG Edan SE 1200 express	0.043	Infinite	100.6	Infinite	Infinite	100.7	0.2	8.8

NC meaning normal conditions, *PE* means protective earth, *AP* applied part, *NE* equipotential node

[a] Denotes devices that did not pass completely or partially the inspection

[b] Denotes measures that were taken on ECG cables as applied parts

TABLE 2.10

The report of the electric safety measurements done in Benin's Hôpital La Croix (H4), Centre Hospitalier Universitaire (CHU) de Zone Suru-Léré (H5), and CHU de Zone d'Abomey-Calavi (H6).

Hospital	Equipment	Protective earth (ohm)	Insulation					Leakage current	
			Mains-PE	AP-PE	Mains-AP	Mains-NE	AP-NE	Equipment	AP
H4	Patient monitor	1.23[a]	Infinite	98.4	Infinite	Infinite	98.2	0.2	7.7
H6	ECG Schiller AT 102	0.126[a]	Infinite	96.2	Infinite	Infinite	96.2	0.1	20.7
	Patient monitor[b]	1.7[a]	Infinite	Infinite	Infinite	Infinite	Infinite	0.5	3.8
	Bionet fetalcare ECG	0.34[a]	Infinite	Infinite	Infinite	Infinite	Infinite	0.65	0.5
H5	Biobase Centrifuge	0.125[a]	Infinite	Infinite	Infinite	Infinite	Infinite	0.5	NA
	Mindray Bs200 Analyzer	0.633[a]	Infinite	Infinite	Infinite	Infinite	Infinite	34.2	NA
	Heamatology analyser	0.073	Infinite	Infinite	Infinite	Infinite	Infinite	0.2	NA
	Sysmex coagulation system	0.166[a]	Infinite	Infinite	Infinite	Infinite	Infinite	0.7	NA
	Patient monitor	NC	Infinite	99.5	Infinite	Infinite	Infinite	0.9	8
	Aspel ECG	NC	Infinite	Infinite	Infinite	Infinite	Infinite	0.2	18.6

NC meaning normal conditions, *PE* means protective earth, *AP* applied part, *NE* equipotential node

[a] Denotes the measures not respecting the standard

[b] Denotes the devices for which no electrodes were available

Discussion and Conclusions

This paper introduces a framework to assess clinical locations, specifically in LMICs, through a semi-structured interview and electrical safety measures. Such a framework, tested in Uganda and Benin, is crucial for mapping out the different realities of healthcare locations of LRSs.

Many developing countries still lack access to MDs and equipment that are appropriate for their specific clinical needs because of poor regulatory controls (36). More importantly, what they do have is often inappropriate due to a mismatch between working conditions and design constraints.

In Africa, for example, competent authorities, who control the quality and safety of MDs, often lack adequately trained staff for consultation (37) due to limited human capacity in biomedical engineering (38).

We believe that the problem of existing standards is that of generalism, or of a non-inclusiveness, which does not take into consideration all the specific realities and that, for this reason, is not reachable by them. This dilemma, however, can be addressed with an inductive method, which, by examining specific situations, can inform the writing of new regulations and standards that are more inclusive and universally applicable. It is only by studying and taking into account different contexts and types of users that the design of a safe and high-quality MD for LRSs can be successful (39). It is true that many challenges are common to these settings, but it is also true that each of them has its own identity and peculiarities. According to this, open-source and collaborative methods have the potential to improve the design of needs-based MDs, offering specific solutions to problems not properly considered by current standards, oriented to well-structured healthcare environments. When properly deployed, the open design paradigm has the potential to increase access to medical technologies, reducing the management, maintenance, and repair costs due to the open-access of device blueprints (40).

In this context, UBORA, the open Biomedical Engineering e-platform for collaborative design, can be effectively used to develop safe and effective MDs, and the relative spare parts (41), the analysis of the technical needs, the risk management process, legal aspects, safety criteria and performance data fundamental for maintaining the compliancy of the repaired devices with the medical device regulation (42).

As shown by the results of the questionnaire, all the analyzed facilities are inefficiently built, with poor isolation from foul odors, smog and undue noises, and unstable and unsafe power supplies. The scarcity of essential MDs, spare parts and consumables, together with a poor maintenance system, bolstered by the chronic lack of biomedical engineers, technicians, and healthcare personnel (from nurses to specialized doctors), hinders the safe and efficient care of patients, impeding universal health coverage.

The electrical safety measurements confirmed the results of the questionnaire, highlighted the common issue of too high protective earths of some MDs, and, above all, clearly proved that in Benin there are problems with the grounding systems.

One limitation of the study is that it was not always possible to test MDs because of the lack of authorization or because they were all being used for the care of patients.

Nevertheless, the availability of the technical staff, who allowed the measurements and the interviews in the field, made us understand the urgency for the locals of the problems highlighted in this paper. Therefore, this study is an important starting point to frame the problem and present the framework which favored a thorough investigation in the field and facilitated the subsequent processing of the data, showing itself a suitable, exportable, and repeatable tool to offer an overview of healthcare locations in specific contexts. The study will be expanded to include more hospitals around Uganda, Benin, and other LMICs in order to provide the basis for promoting awareness of the issues they face and towards global policy changes for health equity.

Note

1 L. Di Pietro, D. Piaggio, I. Oronti, A. Maccaro, R.C. Houosseuvo, D. Medenou, C. De Maria, L. Pecchia, A. Ahluwalia, A Framework for Assessing Healthcare Facilities in Low Resource Settings: Field Studies in Benin and Uganda, in «Journal of Medical and Biological Engineering» (Springer), (24 June 2020) 40:526-534 (DOI: https://doi.org/10.1007/s40846-020-00546-3)

References

(1) United Nations (UN) The sustainable development goals. 2017, March. Available from: https://sustainabledevelopment.un.org/?menu=1300. Accessed 5/5/2020.

(2) Roth L, Bempong D, Babigumira JB, Banoo S, Cooke E, Jeffreys D, et al. Expanding global access to essential medicines: Investment priorities for sustainably strengthening medical product regulatory systems. *Globalization and Health.* 2018;14(1):102.

(3) Chapman AR. Assessing the universal health coverage target in the Sustainable Development Goals from a human rights perspective. *BMC International Health and Human Rights.* 2016;16(1);33.

(4) WHO. *Medical devices: Managing the mismatch: An outcome of the priority medical devices project*: WHO; 2010. Available from: https://apps.who.int/iris/handle/10665/44407. Accessed 5/5/2020.

(5) Piaggio D, Medenou D, Houessouvo RC, Pecchia L. Donation of medical devices in low-income countries: Preliminary results from field studies. In Badnjevic A, Škrbić R, Gurbeta Pokvić L, editors. *International Conference on Medical and Biological Engineering*: Springer; 2019, May, pp. 423–427.

(6) Dyro J., editor. *Clinical engineering handbook*: Elsevier; 2004.

(7) WHO. *Barriers to innovation in the field of medical devices: Background paper 6, August 2010*: WHO; 2010. Available from: https://apps.who.int/iris/handle/10665/70457. Accessed 5/5/2020.

(8) Richards-Kortum R, Oden M. Devices for low-resource health care. *Science.* 2013;*342*(6162);1055–1057.

(9) Malkin RA. Design of health care technologies for the developing world. *Annual Review of Biomedical Engineering.* 2007;9:567–587.

(10) Lustick DR, Zaman MH. Biomedical engineering education and practice challenges and opportunities in improving health in developing countries. In 2011 Atlanta conference on science and innovation policy 2011 Sep 15 (pp. 1–5). IEEE.

(11) Taylor AR, Muelenaer P, Kochersberger KB, Muelenaer AA, Bickford LR, Redican K. Innovating for global health: Study of healthcare technology failure in southern Malawi. *Annals of Global Health*, 2016;82(3):600–601.

(12) Pecchia L, Piaggio D, Maccaro A, Formisano C, Iadanza E. The inadequacy of regulatory frameworks in time of crisis and in low-resource settings: Personal protective equipment and COVID-19. *Health and Technology* 2020 Nov;10(6):1375–83.

(13) Buzdugan MI, Bălan H, Mureşan DT. An electrical power quality problem in an emergency unit from a hospital–Case study. In *SPEEDAM 2010*: IEEE; 2010, June, pp. 251–256.

(14) Rao U, Singh SN, Thakur CK. Power quality issues with medical electronics equipment in hospitals. In *2010 International Conference on Industrial Electronics, Control and Robotics*: IEEE; 2010, December, pp. 34–38.

(15) Malkin RA. Barriers for medical devices for the developing world. *Expert Review of Medical Devices.* 2007;4(6):759–763.

(16) Hsia RY, Mbembati NA, Macfarlane S, Kruk ME. Access to emergency and surgical care in sub-Saharan Africa: The infrastructure gap. *Health Policy and Planning.* 2012;27(3):234–244.

(17) Albutt K, Yorlets RR, Punchak M, Kayima P, Namanya DB, Anderson GA, Shrime MG. You pray to your God: A qualitative analysis of challenges in the provision of safe, timely, and affordable surgical care in Uganda. *PloS One.* 2018 Apr 17;13(4):e0195986.

(18) Dovlo D, Karamagi H. Life-saving hospitals – A role in UHO for Africa. Building health dreams. *World Hospitals and Health Services: The Official Journal of the International Hospital Federation.* 2016; Jan 24;10(1):e0116949.

(19) Murthy S, Leligdowicz A, Adhikari NK. Intensive care unit capacity in low-income countries: A systematic review. *PloS One.* 2015;10(1).

(20) Tumukunde J, Sendagire C, Ttendo SS. Development of intensive care in low-resource regions. *Current Anesthesiology Reports.* 2019;9(1):15–17.

(21) Firth P, Ttendo S. (). Intensive care in low-income countries—A critical need. *New England Journal of Medicine.* 2012;367(21):1974–1976.

(22) Mutale W, Bosomprah S, Shankalala P, Mweemba O, Chilengi R, Kapambwe S, Chishimba C, Mukanu M, Chibutu D, Heimburger D. Assessing capacity and readiness to manage NCDs in primary care setting: Gaps and opportunities based on adapted WHO PEN tool in Zambia. *PloS One.* 2018 Aug 23;13(8):e0200994.

(23) Musinguzi G, Bastiaens H, Wanyenze RK, Mukose A, Van Geertruyden JP, Nuwaha F. Capacity of health facilities to manage hypertension in Mukono and Buikwe districts in Uganda: Challenges and recommendations. *PloS One.* 2015;10(11):e0142312.

(24) United Nations Development Programme (UNDP). *Human Development Report 2019. Beyond income, beyond averages, beyond today: Inequalities in human development in the 21st century*: UNPD; 2019. Available from: http://hdr.undp.org/en/content/human-development-report-2019. Accessed 5/5/2020.
(25) Harrington J. From the Solomon Islands to Liberia: These are the 25 poorest countries in the world. *USA Today*. 2018. Available from: https://eu.usatoday.com/story/money/2018/11/29/poorest-countries-world-2018/38429473/. Accessed 5/5/2020.
(26) United Nations Development Programme. *Global Human Development Indicators 2019*: UNPD; 2019. Available from: http://hdr.undp.org/en/countries/profiles/. Accessed 5/5/2020.
(27) The World Bank. World Bank Open Data. Available from: https://data.worldbank.org/. Accessed 5/5/2020.
(28) The World Bank. Quality of electrical supply. 2017. Available from: https://govdata360.worldbank.org/indicators/hf3350a8e?country=BRA&indicator=41280&viz=line_chart&years=2007,2017. Accessed 5/5/2020.
(29) Central Intelligence Agency. *The World Factbook*: CIA; 2015. Available from: www.cia.gov/library/publications/the-world-factbook/fields/359.html. Accessed 5/5/2020.
(30) Wikipedia contributors. (2019, October 19). Healthcare in Uganda. In *Wikipedia, The Free Encyclopedia*. Retrieved *17*:24, May 5, 2020, from https://en.wikipedia.org/w/index.php?title=Healthcare_in_Uganda&oldid=922014431. Accessed 5/5/2020.
(31) Medenou D, Fagbemi LA, Houessouvo RC, Jossou TR, Ahouandjinou MH, Piaggio D, et al. Medical devices in Sub-Saharan Africa: Optimal assistance via a computerized maintenance management system (CMMS) in Benin. *Health and Technology*. 2019;9(3):219–232.
(32) Singh S. Location & layout of hospital. Master of hospital administration. [Lecture slides]. In press 2008.
(33) AIA Academy of Architecture for Health, & Facilities Guidelines Institute. *Guidelines for design and construction of hospital and health care facilities*: AIA Press; 2006.
(34) WHO. Priority Medical Devices. Available from: www.who.int/medical_devices/priority/en/. Accessed 5/5/2020.
(35) WHO. *Medical equipment maintenance programme overview: WHO Medical device technical series*: WHO; 2011. Available from: www.who.int/medical_devices/publications/medical-equipment-maintenance/en/. Accessed 5/5/2020.
(36) De Maria C, Di Pietro L, Lantada AD, Madete J, Makobore PN, Mridha M, et al. Safe innovation: On medical device legislation in Europe and Africa. *Health Policy and Technology*. 2018;7(2):156–165.
(37) Buckley GJ, Riviere JE, editors. *Ensuring safe foods and medical products through stronger regulatory systems abroad*: National Academies Press; 2012.
(38) Lustick DR, Zaman MH. Biomedical engineering education and practice challenges and opportunities in improving health in developing countries. In *2011 Atlanta Conference on Science and Innovation Policy*: IEEE; 2011, September, pp. 1–5.
(39) De Maria C, Mazzei D, Ahluwalia A. Improving African health care through open source biomedical engineering. *International Journal on Advances in Life Sciences*. 2015;7(1):10–19.
(40) De Maria C, Di Pietro L, Lantada AD, Ravizza A, Mridha M, Torop J, Madete J, Makobore P, Ahluwalia A. The UBORA e-infrastructure for open source innovation in medical technology. In XV Mediterranean Conference on Medical and Biological Engineering and Computing–MEDICON 2019: Proceedings of MEDICON

2019, September 26–28, 2019, Coimbra, Portugal 2020 (pp. 878-882). Springer International Publishing.

(41) Di Pietro L, De Maria C, Ravizza A, Ahluwalia A. Co-design open-source medical devices: How to minimize the human error using UBORA e-infrastructure. In *2019 41st Annual International Conference of the IEEE Engineering in Medicine and Biology Society (EMBC)*: IEEE; 2019, July, pp. 3730–3733.

(42) UBORA. *Euro-Africa Open Biomedical Engineering Innovation e-platform for Innovation through Education*. Available from: https://platform.ubora-biomedical.org/. Accessed 5/5/2020.

Chapter 3

Ethics and Regulation

Personal Protective Equipment and COVID-19: The Inadequacy of Regulatory Frameworks in Time of Crisis and in Low-Resource Settings[1]

Introduction

As of early April 2020, the world is stricken by the recent pandemic outbreak (1) of a new strain of Coronavirus, previously unknown to mankind, denominated Severe Acute Respiratory Syndrome CoronaVirus 2 (SARS-CoV-2). This virus is part of the family of the coronaviruses, which are viruses commonly affecting mammals and birds. Although the respiratory tract infections in humans caused by this family of viruses are usually common colds (Human Coronavirus 229E, Human Coronavirus NL63, Human Coronavirus OC43, and Human Coronavirus HKU1), sometimes they beget viral pneumonia and rarely they can be the cause of a severe acute respiratory syndrome (SARS) (SARS-CoV, MERS-CoV, and SARS-CoV-2) (2). SARS-CoV-2 causes a disease, better known as COVID-19 (CO for Corona, VI for Virus, D for disease and 19 for the year in which it was identified), with symptoms spanning from mild (e.g., fever, tiredness, dry and continuous cough, shortness of breath, diarrhea, and sore throat (3, 4)) to serious (e.g., viral pneumonia and multi-organ failure (5, 6)). SARS-CoV-2, similarly to other viruses (7), seems to have spilled over to humans from wild animals (8). As a consequence, the human immune system, having never been in contact with such a virus, lacks the ability to fight against the pathogen (7), which can have particularly dangerous effects on subjects with already weak immune systems, or immunosuppressed or elderly subjects with existing preconditions. Based on the current information, the virus has been classed as a Hazard Group 3 (HG3) pathogen (9) and the World Health Organization (WHO) has stated that laboratory tests and practices should follow biosafety level 3 guidelines (10).

The COVID19 outbreak's impact on European countries was twofold. First, the supply chain of personal protective equipment (PPE), medical devices (MDs), consumables, and spare parts revealed its frailty in its dependence on China's capability to produce them, severally hindered by the lockdown since January 2020.

DOI: 10.1201/9781003436348-11

Second, this world pandemic has been causing an unprecedented demand of hospitalizations, especially in intensive care units (ICUs), since its early stages. This set off a chain reaction affecting a number of other routine hospitalizations (e.g., elective surgeries), which were postponed giving priority to ICU beds in terms of resources (spaces, personnel, and equipment). Moreover, healthcare staff is highly exposed to the risk of catching COVID-19 themselves due to the inner nature of their daily routine which exposes them to physical contact with patients. The combination of these factors has created in Europe *de facto* conditions that are usually typical of "low-resource settings" (LRSs), generating havoc among all the countries, independently from their wealth status.

The combination of a frail supply chain and an unprecedented demand of ICU beds demonstrated the extent to which countries were not prepared to tackle global disasters, such as the current pandemic.

This is particularly evident in Europe, where many healthcare systems (e.g., France, Italy, Spain, and the United Kingdom), being rated among the best ones in the world (11), are being heavily overburdened by the ever-increasing number of patients needing hospitalization or intensive care (12–15). The national health systems, in fact, lack essential resources[2] for dealing with COVID-19, including MDs (e.g. surgical masks, ventilators, and infusion pumps), PPEs (16), and healthcare personnel (who are being reduced by the disease itself).

For the first time in decades, the progressive scarcity of devices, equipment, and resources has raised also in high-income countries (HICs) the problem of resource allocation and prioritization. The latter could expose a part of the population, probably the most disadvantaged individuals, to further difficulties in accessing healthcare services (17).

The urgent need for equipment directly affects the role of clinical engineers, professionals who are in charge of verifying that all the medical and electro-medical devices are compliant with the essential requirements imposed by the national laws, before authorizing their use in hospital settings. In Europe, this means compliance with the European framework of directives and regulations certified by the presence of the European Conformity (CE) mark. Strictly following international standards is the regular path chosen by the manufacturers in order to guarantee the compliance of their products to the above-mentioned requirements, in terms of performances and safety. The current situation has highlighted the flaws of the regulations. For instance, the non-universality of regulations, norms, and international standards is clearly evident in these situations of emergency. The problem of non-universality of technical norms is well-known in the context of LRSs, especially in the context of low- and middle-income countries (LMICs) as highlighted by the authors of this paper who have been extensively acting to overcome this issue (18, 19). The COVID-19 pandemic is dramatically demonstrating that this limit is paramount also in HICs during emergencies. The international standards, indeed, proved to be often too generic and demanding, resulting difficult to be implemented in many countries, in terms of time, costs, and overall effort required thus jeopardizing a prompt response to emergencies.

This is everyday evidence in lower-income countries, and it is becoming now clear also in high-income ones.

In this critical context, we have joined our efforts to write this manuscript in order to share our considerations on the necessity of identifying a set of minimum requirements to test PPEs for use in hospitals during the COVID-19 pandemic. We hope this contribution may be relevant for readers, helping them navigating the variegated context of PPE's regulatory framework. The proposed approach reflects a minimum set of tests that should always be considered despite the waivers issued by several states. This discussion should then be continued, once this crisis is over, especially with regard to lower-income countries where the inadequacy of international norms is clear also under everyday conditions.

PPE Regulation in EU and Its Inadequacy for COVID-19

The first European directive on the design, manufacture, and marketing of PPE was published in December 1989, exactly 30 years before SARS-CoV-2 was first identified in China (Directive 89/686/EEC). This Directive was superseded in 2016 by Regulation (EU) 2016/425, which outlines legal obligations in order to guarantee the highest quality and efficiency of the PPE circulating in the EU internal market. The compliance with such regulation is assured by the CE marking. As per semantics, the change from directives to regulations was not trivial and implied a tougher and more harmonized approach. Directives, in fact, are EU instruments requiring member states to legislate accordingly and enforce them at a national level through relevant laws with a certain flexibility. *Vice versa*, EU Regulations are simultaneously enforced in each member state after a transitional period, imposing clear and detailed common rules, which do not give room for divergent transpositions by member states (20).

The Regulation (EU) 2016/425 itself refers to harmonized standards, developed by a recognized European Standards Organization (e.g., the European Committee for Standardization [CEN]) as a way in which manufacturers or notified bodies can assess the conformity of a product. As for Personal Eye Protection and Respiratory Protective Devices, EN166:2001 and EN 149:2001+A1:2009 are the respective harmonized standards.

The pathway that manufacturers can take for achieving the CE Marking of PPE are different, depending on the PPE risk category. For PPE used in low-risk conditions, manufacturers can rely on internal labs and tests to demonstrate that their products are compliant with the harmonized standards requirements, issuing self-certifications. When dealing with protection against SARS-CoV-2 virus, the PPE are framed in the highest risk class (class II or III), requiring the involvement of external notified bodies. Notified bodies are appointed under Article 28 of Regulation (EU) 2016/425, and their responsibilities include confirming that PPE have an adequate level of health and safety in accordance with the essential requirements laid down for that product in harmonized standards. According to the number of PPE produced, notified bodies may be requested to repeat this test several times per year. This may include testing of a few samples that manufactures send to the Notified Body, as well as inspections in the manufacturing facilities. A key point is that there is a series of tests that need to be performed according to the relevant standards, depending on the type of device and equipment, to prove that the designed solutions fulfil the minimum requirements in terms of performance and safety. This series of tests is not specific for hospital settings, but it is general, aiming at including all the working conditions to which PPE could be exposed, including testing the robustness of PPE after having been exposed to extreme temperatures (e.g., 70°C), which are never reached in hospitals due to harmonized standards for medical locations.

Therefore, when dealing with general standards, their generality becomes a too stringent constraint in emergency situations. For example, PPE such as face shields or particle filtering face masks (e.g., FFP2 or FFP3 masks) are also (and more often) required in settings other than hospitals, such as in carpentry or soldering activities, in conditions very different from the ones needed by hospital settings. The tests required by the international norms reflect these extreme working conditions, which are different from the ones that can be found in hospitals. Some national standards developing organizations and some notified bodies, aware of the fact that requiring redundant tests is not strictly necessary in conditions of emergency, reconsidered the whole procedure for testing PPE and reduced it to a subset of essential tests and minimum requirements that should be met for use in a COVID-19 hospital department.

Confirming this criticism, on the 13 March 2020, the European Commission published the Recommendation EU 2020/403 (non-legislative act), providing guidance for conformity assessment and market surveillance procedures within the context of the COVID-19 threat. This recommendation has proved the resiliency of the European Commission and its outstanding capability to react to crisis, while confirming the inadequacy of European Regulation for crises and scarce resource scenarios.

Amid this widespread confusion, the scientific community has to make clear the fact that, especially during a crisis, PPE should meet the highest possible quality standards, in accordance to the ALARP (As Low As Reasonably Possible) risk management principle (21).

Amateur Solutions and the Importance of International Standards and CE Marking

The importance of meeting high-quality standards, guaranteeing the efficiency of MDs and equipment and the safety of their users, is evident now more than ever. During this pandemic, we are witnessing growing proliferation of amateur initiatives (i.e., quick fixes) aiming at providing for the above-mentioned needs. These initiatives, amplified through social media, although certainly driven by good intentions, may beget a series of solutions potentially as harmful as the problem they are trying to solve if not properly mentored. The innumerable solutions one can come across on the internet span from using baking paper to reproduce a (surgical!) face mask to 3D printing filtering facial masks using cotton filters, claiming that they are effective in filtering the virus. The lack of any risk assessment, albeit minimal, poses major risks for the user. Using materials from vacuum cleaner filters to realize filtering face masks, for example, could be a threat for the user's safety, in case of the presence of dangerous glass microfibers (22).

Also, in this context, regulations and standards are essential, as they sum up the state of the art resulting from a series of field experiences aimed at guiding the manufacturer in designing and producing devices with high levels of safety. In this regard, visual inspections performed by experienced technicians are the first approaches that can be used to evaluate if the obtained prototype is safe by design. For example, even a small abrasion or defect due to suboptimal materials or inappropriate manufacturing processes can potentially lead to discomfort and skin irritation or lesions in the long run. This possibility, potentially dangerous in normal working conditions, is even more risky in extraordinary working conditions. In severe working conditions (due to extended/more frequent shifts for lack of personnel and stressful conditions) mistakes or distractions are even more likely. A robust design is thought to be resilient to these conditions as well.

The Review of Standards in Situations of Emergency – a Case Study in the United Kingdom

The United Kingdom (UK) Office for Product Safety and Standards (OPSS) has been working with manufacturers to understand where regulatory requirements were preventing them from delivering the products the public and National Health System (NHS) needed, offering continuously updates and guidance.

On 10 April 2020, the OPSS position is that manufacturers can deploy PPE provided that:

- It meets the essential safety requirements
- Conformity assessment procedures have been started via a Notified Body, even if the conformity assessment, "including affixing of CE marking procedures", has not been completed.

The OPSS has also issued an official letter to UK notified bodies providing further guidance on the European Commission Recommendation (EU 2020/403) and urging them to prioritize the conformity assessment of PPE necessary for protection in the context of COVID-19, and requesting this work to be conducted swiftly. With this letter, notified bodies were also allowed to test the products using other than harmonized standards, nominally WHO recommendations.

The UK notified bodies reacted promptly, prioritizing such tests aiming at assessing essential safety requirements specifically thinking of NHS workers during the COVID-19 pandemic. As a result, PPE can now be tested much faster (e.g., in one week for face shields) for one third of the usual cost *circa* and with also a reduced number of samples to be sent for the tests. Certainly, this has been a prompt and functional response. However, once this COVID-19 crisis is overcome, there should be a wider discussion on how to make norms and regulations intrinsically more flexible so as to always allow rapid interventions to ever-changing situations and to be inclusive of many different specificities. This could potentially be achieved either by including some default waivers which could allow modifications in case of need, or by reformulating the way the regulations are conceived, basing the new ones on a better defined and effective universality.

Face Shields

Face shields (aka visors) are part of a family of devices named personal eye protective equipment, as defined in EN 166:2001. According to EN ISO 4007:2018, the term "protector" (3.5.1.1) for purposes of eye and face protection includes eye protectors (3.5.1.2), eye guards (3.5.1.4), face protectors (3.5.1.5), face screens/shields (3.5.1.6), and more (i.e. goggles, hand shields, protective masks, spectacles, visors).

As aforementioned EN 166 provides for a series of tests for the evaluation of face shields that are inclusive of any possible future use, including the industrial environment in which the worker is at high risk of injuring his/her eyes and/or face with splinters or electric arcs. However, to respond to the urgent needs related to COVID-19, national standards bodies have reduced the number and type of essential tests to be performed on face shields to be used in healthcare settings for COVID-19. This fits perfectly with the concept of norms as technical means to reach the state of the art. Technical standards are never mandatory. The authors of this paper have been in touch with different international experts and standards bodies and drafted their own subset of suggested tests. Table 3.1 summarizes the tests that the authors of this paper believe to be essential for face shields that are to be used as an additional protection by healthcare workers in this pandemic. EN 166 refers to EN 167:2002 for some test specifications; however, EN167 was recently superseded by EN ISO 18526:2020. Therefore, the authors created Table 3.1 inserting the links to the correct clauses in the new standard, where applicable.

Filtering Half Masks

Filtering half masks to protect against particles are part of a family of equipment named "respiratory protective devices", which are regulated by EN 149:2001+A1:2009. They are defined as respiratory protective devices that consist substantially of filter material or comprise a face piece in which the main filter(s) form an inseparable part of the

TABLE 3.1

The subset of tests (EN 166) deemed essential to evaluate personal eye protection equipment for COVID-19.

Performance requirement	Test method clause	Requirement
General construction (Section 6.1 EN 166:2002)	Visual inspection and manufacturer's certificates	Eye-protectors shall be free from projections, sharp edges, or other defects which are likely to cause discomfort or injury during use.
Materials (6.2)	Visual inspection and manufacturer's certificates	No parts of the eye-protector which are in contact with the wearer shall be made of materials which are known to cause any skin irritation.
Headbands (6.3)	By measuring	Headbands, when used as the principal means of retention, shall be at least 10 mm wide over any portion which may come into contact with the wearer's head. Headbands shall be adjustable or self-adjusting.
Field of vision (7.1.1)	The size of the field of vision is defined in conjunction with the appropriate head-form described **in clause 17 of EN 168:2001.** The test shall be carried out in accordance **with clause 18 of EN 168:2001.**	Eye-protectors shall exhibit a minimum field of vision defined by the two ellipses in Figure 1 when placed and centred at a distance of 25 mm from the surface of the eyes of the appropriate head-form. The horizontal axis shall be parallel to and 0,7 mm below the height of the line connecting the centres of the two eyes. The horizontal length of the ellipses shall be of 22.0 mm, the vertical width of the ellipses shall be 20.0 mm. The centre distance of the two ellipses shall be d = c + 6 mm, where c is the pupillary distance. The pupillary distance is 64 mm for the medium head-form and 54 mm for the small head-form, if not specified differently by the manufacture.
Spherical, astigmatic and prismatic refractive powers (7.1.2.1)	The refractive powers of oculars shall be measured by the reference methods specified **in clause 6 of EN ISO 18526-1:2020.** If, during measurement using the telescope, a doubling or other aberration of the image is observed, then the test sample shall be subjected to further examination using the test method described in **6.3 of EN ISO 18526-1:2020**	The permissible tolerances for oculars without corrective effect are given in Table 3 of **Clause 7.1.2.1.2 of EN166**.

(*continued*)

TABLE 3.1 (Continued)

The subset of tests (EN 166) deemed essential to evaluate personal eye protection equipment for COVID-19.

Performance requirement	Test method clause	Requirement
Transmittance – oculars without filtering action (7.1.2.2-1)	**Clause 6 of EN ISO 18526-2:2020**	Oculars intended to protect the eyes against mechanical or chemical hazards only, and cover plates, shall have a luminous transmittance greater than 74.4%.
Diffusion of light (7.1.2.3)	**Clause 14 of EN ISO 18526-2:2020**	The max value of the reduced luminance factor shall be 0.50 cd/(m^2*lx) for all the other oculars.
Quality of material and surface (7.1.3)	**Clause 6.6 of EN ISO 18526-3:2020**	Except for a marginal area 5 mm wide, oculars shall be free from any significant defects likely to impair vision in use, such as bubbles, scratches, inclusions, dull spots, pitting, mould marks, scouring, grains, pocking, scaling and undulation.
Minimum robustness (7.1.4.1)	**Clause 3.1 of EN 168:2001**	The 7.1.4.1 requires performing an impact test with a 22 mm nominal diameter steel ball, impacting with a force of (100 ± 2) N. On so testing the following defects shall not occur: a) Ocular fracture: an ocular shall be considered to have fractured if it cracks through its entire thickness into two or more pieces, or if more than 5 mg of the ocular material becomes detached from the surface away from the one in contact with the ball, or if the ball passes through the ocular; b) Ocular deformation: an ocular shall be considered to have been deformed if a mark appears on the white paper on the opposite side to the one on which the force is applied. Note: in period of crisis such as COVID, this should be at least tested using realistic simulations
Resistance to corrosion (7.1.6)	**Clause 6.9 of EN ISO 18526-3:2020**	After the test, all metal parts of the eye-protector shall display smooth surfaces, free from corrosion, when they are examined by a trained observer. Note: This should be extended to plastic parts considering the sterilization processes and substances (e.g., after cleaning visibility should not be compromised)
Resistance to ignition (7.1.7)	**Clause 6.10 of EN ISO 18526-3:2020**	Eye-protectors shall be considered to be satisfactory if no part of the eye-protector ignites or continues to glow after removal of the steel rod.

TABLE 3.1 (Continued)

The subset of tests (EN 166) deemed essential to evaluate personal eye protection equipment for COVID-19.

Performance requirement	Test method clause	Requirement
Protection against (droplets) and splashes of liquid (7.2.4)	**Clause 12 of EN 168:2001**	Face shields cover the eye-region rectangle of the appropriate head-form as described in 10.2.2.2 of EN 168:2001 as assessed in accordance with 10.2 of EN 168:2001. Additionally, face-shields for protection against splashes of liquids shall have a viewing area with a minimum vertical centre-line depth of 150 mm when mounted in the appropriate housing.
Lateral protection (7.2.8)	**Clause 6.4 of EN ISO 18526-3:2020**	Eye-protectors claimed to provide lateral protection shall pass the lateral region coverage assessment
Resistance to fogging (7.3.2)	**Clause 6.11 of EN ISO 18526-3:2020**	If oculars are described as resistant to fogging they shall remain free from fogging for a minimum of 8 s when tested.

device. These devices can be provided with valves for inspiration and exhalation. Depending on their filtering power, they can be designated as FFP1, FFP2, and FFP3. FFP2 and FFP3 half masks, which are suggested as a protection against COVID-19 (23), belong to category III and require, for CE marking, a certificate from and external Notified Body, issued after exams on the product's technical documentation and tests on the product itself, as stated in the European Regulation 2016/425.

As per the face shields, the tests enlisted for these devices are also designed taking into account all the most disparate conditions of use. The authors of this paper, following discussions with different national standards bodies, summarized in Table 3.2 the tests that they believe to be essential for this piece of equipment to be used by healthcare workers in this pandemic.

Similarities Between the COVID-19 Pandemic and Low-Resource Settings

COVID-19 pandemic and LRSs have deep differences, but also many commonalities. Low-resource settings are common in LMICs, but also exist in rural and remote areas of many high-income ones. Contrary to the common belief, the main problem of LRSs lays beyond the lack of funding. Several studies have, indeed, highlighted the fact that if this had been the main problem, donations would have solved it (19, 24). Conversely, modern medicine requires much more than budget, as is evident

TABLE 3.2

The subset of tests (EN 149:2001+A1:2009) deemed essential to evaluate respiratory protective devices for COVID-19. The underlining highlights the parts of the test method clauses that were changed from the original document.

Performance requirement	Test method clause	Requirement
Visual inspection (Section 7.3 EN 149:2001+ A1:2009)	The visual inspection is carried out where appropriate by the test house prior to laboratory or practical performance tests (**Clause 8.2**).	
Packaging (7.4)	The visual inspection is carried out where appropriate by the test house prior to laboratory or practical performance tests (**Clause 8.2**).	Particle filtering half masks shall be offered for sale packaged in such a way that they are protected against mechanical damage and contamination before use.
Cleaning and disinfecting (7.6) If the particle filtering half mask is designed to be re-usable, the materials used shall withstand the cleaning and disinfecting agents and procedures to be specified by the manufacturer.	Testing shall be done in accordance with **Clause 8.4 and Clause 8.5**. With reference to 7.9.2, after cleaning and disinfecting the re-usable particle filtering half mask shall satisfy the penetration requirement of the relevant class. Testing shall be done in accordance with **Clause 8.11**.	After cleaning and disinfecting the re-usable particle filtering half mask shall satisfy the penetration requirement of the relevant class.
Practical performance (7.7) The particle filtering half mask shall undergo practical performance tests under realistic conditions. These general tests serve the purpose of checking the equipment for imperfections that cannot be determined by the tests described elsewhere in this standard. Where practical performance tests show the apparatus has imperfections related to wearer's acceptance, the test house shall provide full details of those parts of the practical performance tests which revealed these imperfections.	2 particle filtering half masks shall be tested as received in accordance with **Clause 8.4**.	During the tests the particle filtering half mask shall be subjectively assessed by the wearer and after the test, comments on the following shall be recorded: a) head harness comfort; b) security of fastenings; c) field of vision; d) any other comments reported by the wearer on request.

TABLE 3.2 (Continued)

The subset of tests (EN 149:2001+A1:2009) deemed essential to evaluate respiratory protective devices for COVID-19. The underlining highlights the parts of the test method clauses that were changed from the original document.

Performance requirement	Test method clause	Requirement
Total inward leakage (7.9.1) The laboratory tests shall indicate that the particle filtering half mask can be used by the wearer to protect with high probability against the potential hazard to be expected. The total inward leakage consists of three components: face seal leakage, exhalation valve leakage (if exhalation valve fitted) and filter penetration.	At least 5 specimens shall be tested as received in accordance with **Clause 8.5**.	All samples must achieve the specifics. All individual exercise results tests shall be not greater than 11% (FFP2) (5% for FPP3) and, in addition, all arithmetic means for the total inward leakage shall be not greater than 8% (FFP2) (2% for FFP3).
Penetration of filter material (7.9.2)	At least 3 samples shall be tested as received for NaCl and paraffin oil (PO) for 3 min in accordance with **Clause 8.11.**	The maximum penetration of test aerosol shall be: • 6% for both PO and NaCl for FFP2 • 1% for both PO and NaCl for FFP3
Carbon dioxide content of the inhalation air (7.12)	At least 3 particle filtering half masks shall be tested in accordance with **Clause 8.7**.	The carbon dioxide content of the inhalation air (dead space) shall not exceed an average of 1.0 % (by volume).
Head harness (7.13)	Testing shall be done in accordance with **Clause 8.4 and 8.5**.	The head harness shall be designed so that the particle filtering half mask can be donned and removed easily. The head harness shall be adjustable or self-adjusting and shall be sufficiently robust to hold the particle filtering half mask firmly in position and be capable of maintaining total inward leakage requirements for the device.
Field of vision (7.14)	Testing shall be done in accordance with **Clause 8.4.**	The field of vision is acceptable if determined so in practical performance tests.
Breathing resistance (7.16)	At least 3 samples shall be tested as received in accordance with **Clause 8.9.**	The maximum permitted resistances per volumetric flow rate for FFP2 are: • 30 l/min – 0.7mbar (inhale) • 95 l/min – 2.4mbar (inhale) • 160 l/min – 3.0mbar (exhale) For FFP3 are: • 30 l/min – 1.0mbar (inhale) • 95 l/min – 3.0mbar (inhale) • 160 l/min – 3.0mbar (exhale)

especially in the remit of MDs and equipment, included PPE. In fact, LRSs are often characterized by the lack of clinical knowledge, lack of specialized clinical personnel and technical staff, scarcity of medical devices, drugs, and spare parts due to a jeopardized supply chain (19, 25). As argued above, COVID-19 has simultaneously hindered the supply chain, increased the ICU hospitalization demand, and reduced staff. This created *de facto* conditions that are quite common in LRSs, especially in LMICs.

The COVID-19 pandemic demonstrated the lack of knowledge and lack of preparedness of many HICs, aggravated by the slowness in perceiving the complexity of the situation faced by other countries affected by COVID-19 months in advance. For instance, before facing the COVID-19 outbreak in Milan metropolitan area, Italian authorities and experts of virology failed to understand the complexity of the situation in China and failed to acknowledge the great work done by Chinese colleagues. The same inertia has affected many Northern European countries, which failed to appreciate the complexity of the COVID-19 disaster in Italy, despite the prompt response of Regional Institutions, especially in the South. While we write this paper, it seems that the United States of America's (USA) reaction is again demonstrating some degree of inertia, failing again to acknowledge the severity of the situation in many European countries, despite the effort that those countries are putting into the limiting of this pandemic.

Despite the deep differences, the shift in methodology from LRSs to the COVID-19 pandemic response may help in speeding-up the response to the emergency, also in HICs. After all, shifting from domains where we lack solutions to a domain where we have established methods and tools, has been one of the most powerful engineering solutions. An example could be the application of transformations such as Laplace transform to differential problems, bringing them in a domain where the equations become algebraic and can be easily solved in analytic form, and then shifted in the original domain using an anti-transform.

Conclusion

The COVID-19 outbreak has shown clearly the unsuitability of PPE's regulatory framework, body of norms, and international standards to extreme conditions. This was evident to the professionals working in LRSs, such as LMICs, and it has also now powerfully emerged for HICs during the COVID-19 pandemic. The European regulatory framework evolved in the nineties, mainly to protect European manufacturers from the unsustainable competition from manufacturers producing abroad. This evolution has also been driven by the manufacturers' need to produce PPE for the widest market possible, therefore following the principle of generalism (i.e., PPE tested to be used in any context) as opposed to particularism (i.e., PPE tested to be used in a specific context, such as by nurses working in hospital wards). The prevalence of generalism over particularism resulted in a loss of universality, and in the fact that norms that can be sustained in normal conditions, at least by HICs, become unsustainable in times of crisis. These norms, which are also often assumed as *de facto* standards in many non-EU countries (e.g., in many African countries), are clearly not sufficiently universal for the contexts of LMICs.

In this manuscript, two examples of simplified protocols starting from existing harmonized norms are presented and discussed. Similar exemplifications are currently accepted by European notified bodies in some EU countries and could guide the realization of tests in LMICs. Starting from this unprecedented crisis, HICs will have to reconsider the nature of this regulatory framework and of these norms and international standards. The main lessons that our community should learn from this terrible experience is that there is a major need for an evidence-based regulatory framework, responding to the need of lead and lay users, rather than those of the market itself.

Notes

1 L. Pecchia, D. Piaggio, A. Maccaro, C. Formisano, E. Iadanza, The inadequacy of regulatory frameworks in time of crisis and in Low-Resource Settings: Personal Protective Equipment and COVID19, in «Health and Technology» (Springer), 2 May 2020:https://link.springer.com/article/10.1007/s12553-020-00429-2 (DOI: https://doi.org/10.1007/s12553-020-00429-2)

2 www.who.int/emergencies/diseases/novel-coronavirus-2019/technical-guidance/covid-19-critical-items

References

(1) WHO. WHO announces COVID-19 outbreak a pandemic. 2020. Available from: www.euro.who.int/en/health-topics/health-emergencies/coronavirus-covid-19/news/news/2020/3/who-announces-covid-19-outbreak-a-pandemic

(2) King AM, Lefkowitz E, Adams MJ, Carstens EB. *Virus taxonomy: Ninth report of the International Committee on Taxonomy of Viruses*: Elsevier; 2011.

(3) NHS. Symptoms and what to do – Coronavirus (COVID-19). 2020. Available from: www.nhs.uk/conditions/coronavirus-covid-19/symptoms-and-what-to-do/

(4) WHO. Coronavirus. 2020. Available from: www.who.int/health-topics/coronavirus#tab=tab_3

(5) Tian S, Hu W, Niu L, Liu H, Xu H, Xiao S-Y. Pulmonary pathology of early phase 2019 novel coronavirus (COVID-19) pneumonia in two patients with lung cancer. *Journal of Thoracic Oncology*. 2020 May 1;15(5):700–4.

(6) Wang T, Du Z, Zhu F, Cao Z, An Y, Gao Y, Jiang B. Comorbidities and multi-organ injuries in the treatment of COVID-19. *The Lancet*. 2020 Mar 21;395(10228):e52.

(7) Mandl JN, Schneider C, Schneider DS, Baker ML. Going to bat (s) for studies of disease tolerance. *Frontiers in Immunology*. 2018;9:2112.

(8) Rodriguez-Morales AJ, Bonilla-Aldana DK, Balbin-Ramon GJ, Rabaan AA, Sah R, Paniz-Mondolfi A, et al. History is repeating itself: Probable zoonotic spillover as the cause of the 2019 novel coronavirus epidemic. *Infez Med*. 2020;28(1):3–5.

(9) PHE. COVID-19: Safe handling and processing for samples in laboratories. 2020. Available from: www.gov.uk/government/publications/wuhan-novel-coronavirus-guidance-for-clinical-diagnostic-laboratories/wuhan-novel-coronavirus-handling-and-processing-of-laboratory-specimens#fnref:3

(10) WHO. Laboratory biosafety guidance related to the novel coronavirus (2019-nCoV). 2020. Available from: www.who.int/docs/default-source/coronaviruse/laboratory-biosafety-novel-coronavirus-version-1-1.pdf?sfvrsn=912a9847_2

(11) Worldpopulationreview. Best healthcare in the world. 2020. Available from: https://worldpopulationreview.com/countries/best-healthcare-in-the-world/

(12) Horton R. Offline: COVID-19 and the NHS—"A national scandal". *The Lancet*. 2020;395(10229):1022.

(13) Legido-Quigley H, Mateos-García JT, Campos VR, Gea-Sánchez M, Muntaner C, McKee M. The resilience of the Spanish health system against the COVID-19 pandemic. *The Lancet Public Health*. 2020 May 1;5(5):e251–2.

(14) Remuzzi A, Remuzzi G. COVID-19 and Italy: What next? *The Lancet*. 2020 Apr 11;395(10231):1225–8.

(15) Borges A. French health system under pressure with COVID-19 cases still soaring. 2020. Available from: www.euronews.com/2020/03/17/french-health-system-under-pressure-as-covid-19-cases-still-soaring

(16) Mahase E. Covid-19: Hoarding and misuse of protective gear is jeopardising the response, WHO warns. *British Medical Journal*. 2020 Mar 4;368:m869.

(17) Gostin LO, Friedman EA, Wetter SA. Responding to COVID-19: How to navigate a public health emergency legally and ethically. *Hastings Center Report*. 2020 Mar;50(2):8–12.

(18) Pecchia L. *Health technology assessment of medical devices in low and middle income countries: Study design and preliminary results*. EMBEC & NBC 2017: Springer; 2017. p. 225–258.

(19) Piaggio D, Medenou D, Houessouvo RC, Pecchia L, editors. Donation of medical devices in low-income countries: preliminary results from field studies. In CMBEBIH 2019: Proceedings of the International Conference on Medical and Biological Engineering, 16–18 May 2019, Banja Luka, Bosnia and Herzegovina 2020 (pp. 423–427). Springer International Publishing.

(20) Steiner J, Woods L, Twigg-Flesner C. *EU law*. 9th ed.: Oxford University Press; 2006.

(21) Alexander K, Clarkson P. Good design practice for medical devices and equipment, Part I: A review of current literature. *Journal of Medical Engineering & Technology*. 2000;24(1):5–13.

(22) Sripaiboonkij P, Sripaiboonkij N, Phanprasit W, Jaakkola MS. Respiratory and skin health among glass microfiber production workers: A cross-sectional study. *Environmental Health*. 2009;8(1):36.

(23) Repici A, Maselli R, Colombo M, Gabbiadini R, Spadaccini M, Anderloni A, Carrara S, Fugazza A, Di Leo M, Galtieri PA, Pellegatta G. Coronavirus (COVID-19) outbreak: What the department of endoscopy should know. *Gastrointestinal Endoscopy*. 2020 Jul 1;92(1):192–7.

(24) Williams DB, Kohler JC. Maximizing the value of donated medical equipment in resource-limited settings: The roles of donors and end-users. *CMBES Proceedings*. 2016 May 24;39.

(25) Medenou D, Fagbemi LA, Houessouvo RC, Jossou TR, Ahouandjinou MH, Piaggio D, et al. Medical devices in sub-Saharan Africa: optimal assistance via a computerized maintenance management system (CMMS) in Benin. *Health and Technology*. 2019 May 19;9:219-32.

On the Universality of Medical Device Regulations: The Case of Benin[1]

Background

Medical device regulations and standards

Medical device regulations (MDRs) are essential for improving public health outcomes and increasing access to safe, efficient, effective, and quality medical products (1). The need for MDRs started surfacing in the early 1960s due to some scandals in the "twin" industry of pharma, and was pushed forward by other scandals within the medical device industry (e.g., that of Poly Implant Prothèse [PIP] breast implants) (2). Until the 1970s, the existing regulatory frameworks were based predominantly at the national level, following a prescriptive and subjective approach (3). Internationally, the situation changed with the Medical Device Amendments (United States of America [USA], 1976), that aimed to assure the safety and effectiveness of medical devices (MDs), introducing a risk-based MD classification and establishing the regulatory pathways for new MDs, as well as the post-market requirements (4). These Amendments laid the basis for the European "New Approach", which would constitute the legal foundation of the MD framework of the 1990s, i.e., the Active Implantable Medical Device Directive 90/385/EEC (5) and the Medical Device Directive 93/42/EEC (6), and of the early 2000s Japanese Pharmaceutical Affairs Law.

Despite an ongoing effort for shifting towards more and more harmonized regulations (e.g., the Global Harmonization Task Force and its successor, i.e., the International Medical Device Regulators Forum[2]), currently, globally, the situation is still fragmented: in fact, the three main existing MDRs are the USA Food and Drug Administration (FDA) MDR, the European MDR 2017/745, and the Japanese Pharmaceutical Affairs Law. All these are currently considered the most relevant ones, because over 75% of the MD market is ruled by these three high-income countries (HICs) (7). Nonetheless, the MD market shares of these three countries have been diminishing since 2007 (90% of the global MD) (8). This is due to the emergence of novel, fast-growing MD markets, e.g., China, Canada, Brazil, and India (7).

DOI: 10.1201/9781003436348-12

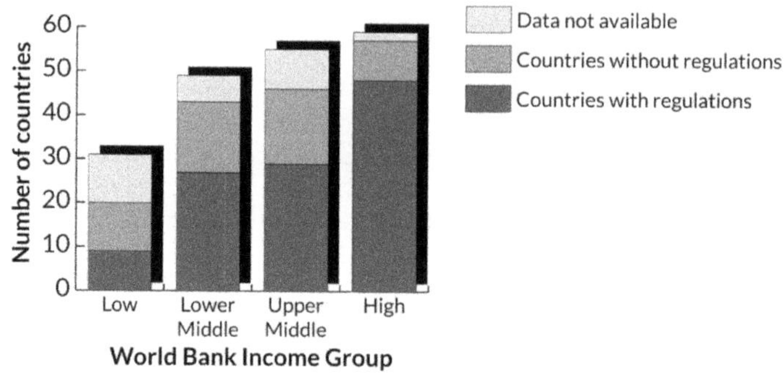

FIGURE 3.1
Number of countries with a legal framework for MDs by income group. Adapted from (12).

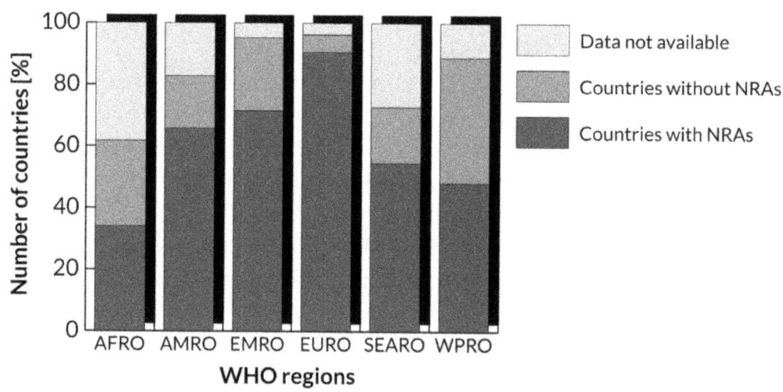

FIGURE 3.2
Existence of NRAs by WHO region. Adapted from (12).

Also Africa, with a compound annual growth rate of about 6% (9), can be included among these novel markets. However, when it comes to regulatory frameworks, there is still a chasm between low- and middle-income countries (LMICs) and HICs. The latter can, in fact, safeguard the safety and efficiency of MDs relying on the presence of very strict MDRs and of national regulatory agencies (NRAs), when compared to lower-income countries (see Figures 3.1 and 3.2) (10). Among the World Health Organization (WHO) regions, the AFRO region is the one with the greatest share of countries without NRAs along with WPRO, while EURO and AMRO are the one with the least share. National regulatory agencies are vital because they are in charge of ensuring that the products that are released for public distribution, including MDs, have been thoroughly reviewed and fulfil the international requirements for safety and quality.

In particular, MDRs focus on three phases of the MD lifecycle, i.e., pre-market, placing on the market, and post-market regulations (10). As regards pre-market regulations, for many LMICs there is not an official definition of MD, of the MD risk classes, nor of their essential principles. Similarly, concerning the placing on the market, several LMICs are lacking an official registration of establishments and listings of MDs and import controls. Lastly, as regards post-market regulations, LMICs have inadequate (if existent) adverse event reporting systems. In respect to this, the Global Atlas of MDs (10) shows that the African Region is the most challenging one regarding the availability of MD regulatory frameworks or data.

A brief summary of the regulatory situation in other world regions follows (11):

Asia: Big markets such as China, Russia, and Japan have strict MDRs in place; smaller markets in the southeast, belonging to the Association of Southeast Asian Nations, agreed on a MD directive.

America (not including the USA): In North America, Canada has its own regulation, issued by Health Canada. In South America, there are also strict regulatory systems.

Africa: Given the wide range of different and complex realities, economic, political, and social instabilities, MDRs in Africa are not well defined, unless the scope of the device is related to the treatment of a specific infectious disease such as malaria, AIDS, and tuberculosis. In such cases, regulations may be present and strengthened by the national regulatory authorities with the aid of help organizations. Further information on the fragmented situation in Africa follows:

Countries with pre-market, placing on the market, and post-market elements: Morocco, Sierra Leone, Burkina Faso, Ghana, Nigeria, Ethiopia, Kenya, and South Africa.

Countries with pre-market and placing on the market elements: Algeria, Sudan, Uganda, Rwanda, Tanzania, Zambia, and Cape Verde.

Countries with pre-market and post-market elements: Egypt and Togo.

Countries with placing on the market elements: Zimbabwe.

Countries without any type of element: Guinea Bissau, Senegal, The Gambia, Mali, Niger, Chad, Libya, Tunisia, Djibouti, Somalia, Ivory Coast, Gabon, Cameroon, Central Republic of Africa, Republic of Congo, Angola, Namibia, Botswana, Lesotho, Eswatini, Burundi, Seychelles, and Madagascar.

Countries with no data available: Mauritania, Guinea-Conakry, Liberia, Benin, South Sudan, Congo Kinshasa, Eritrea, Malawi, Mozambique, Equatorial Guinea, São Tomé and Príncipe.

Oceania: New Zealand and Australia have solid regulations in place, under MedSafe and the Australian Therapeutic Goods Administration.

A Focus on the West African Economic and Monetary Union (UEMOA)

As this article will focus on the case of Benin, after an overview of the global situation, this section will briefly explain the situation in the West African Economic and Monetary Union (UEMOA), of which Benin is a member. The UEMOA is an international organization of eight West African countries[3], created to promote economic integration. This part of Africa was colonized by England, Spain, France, Portugal, and Germany. Therefore, the organization of its healthcare systems is similar to the one established during the colonial period (12). Beyond economic integration, the harmonization of the regulatory frameworks for drugs and MDs was one of the reasons behind the creation of this union. However, no novel community text regarding this has been published since 2010 (13). In particular, the latest regulation concerning the harmonization of pharmaceutical regulation within the UEMOA dates back to 2005 (Reglement N° 02/2005/CM/UEMOA). However, in 2019, the health ministries of the UEMOA reconvened to bring this harmonization forward[4].

Aims and Scope

This article is innovative in the way it addresses more than one global challenge, i.e., the 3rd United Nations Sustainable Development Goal (SDG3: Good health and well-being), and the 10th (SDG10: Reducing inequalities), combining multi- and interdisciplinary knowledge and methods (i.e., biomedical engineer, ethics, sociology, and law), outclassing the Cartesian fragmentation of knowledge (14). This will allow the authors to understand and approach the field from the operational perspective of participatory research. In fact, the article offers an interpretative and operative strategy through the proposed hermeneutic heuristic framework. Dissociating from multiculturalism and its concepts of the non-integrability of cultures, such framework builds on the theory of interculturalism that allows and recognizes the possibility of free cultural and political exchange among countries and cultures without bias[5], and proposes new strategies for an inclusive future.

The authors started their research from the surfacing gap between the regulations on MDs and their contextual application in low-resource settings (LRSs) (16) and addressed this question via eight field studies from 2016 to 2019 across Sub-Saharan Africa (specifically in Benin, Uganda, Ethiopia, and South Africa). Throughout these field studies, the triangulation of qualitative (led by experts in sociology, bioethics, and law) and quantitative methods (led by experts in engineering) was used. In fact, the former can complement the latter when used as a crucial initial step to quantitative research and can also be supplementary information to further validate the quantitative results (17, 18). Such methods include focus groups, content analysis for drafting and validating the questionnaire, pilot tests, semi-structured interviews, medical locations, and MD inspections. The last field study was carried out in November 2019 in Benin, when the multidisciplinary team conducted interviews, surveys, and technical assessments of healthcare locations. The results of the technical part, aimed at assessing the local condition of healthcare settings and MDs, as well as creating a framework for the design of MDs resilient to LRSs, have already been published (17, 19). This article will report the results of the sociological part of the research aimed at scoping local needs and perspectives in respect to regulations, focusing on the case of Benin. The article will also feature a critical discussion of the results, proposing a framework that enables MDs designers and policy makers to address local needs, and giving a crucial overview from the point of view of international law.

Methodology

Given the aforementioned purposes, the authors decided to take into consideration and analyze the case of Benin, which can be an excellent case study, as this article will demonstrate. Benin, in fact, does not have a proper MDR in place, and mainly refers to other existing ones, namely the European Regulation on MDs 745/2017, as Europe's vestiges of colonialism – Benin was a French colony from the late 19th century until 1960 – persist in the way that Europe remains a legal point of reference, as well as other regulations based on the provenience of the acquired or donated MDs (20). Furthermore, Benin is exemplar for evaluating the local and cultural acceptance of MDs or, more generally, healthcare technologies, as it is considered the birthplace of the Vodun religion and of a well-spread, vivid animistic culture. This peculiar situation inspired the authors to create a multidisciplinary research group, which leverages previous collaboration. In particular, scientific collaboration was instituted among the Beninese Děpartment du Génie Biomédical (GBM) de l'Ecole Polytechnique d'Abomey-Calavi (EPAC) and the Laboratoire de Anthropologie Médicale Appliqué (LAMA) of the same university, and the Applied Biomedical Signal Processing and Intelligent e-Health (ABSPIE) Lab and the Warwick Interdisciplinary Research Centre for International Development (WICID) of the University of Warwick.

The early phase of this work began in October 2019, when, in light of the upcoming field study in Benin, some of the authors (A.M., D.P., L.P.) met some of the Beninese members of the research group at the third International Clinical Engineering and Health Technology Management Congress (ICEHTMC III) (Rome, 20-23 October 2019). On this occasion, apart from their presentations at the conference[6], they had the opportunity to listen the WHO's experts stress the importance of inclusive regulations[7] . Moreover, at the same event, focus groups were held with multidisciplinary experts to highlight the themes of interest that would then be the basis for the semi-structured interviews of the upcoming field study (the full questionnaire can be found in Additional material 1; see Additional material 2 to view summary statistics of some selected interview extracts). The themes being:

MDRs in Low-Resource Setting (LRSs)

The role of MD maintenance in LRSs

Beninese rural population perception on MDs and health technologies

Based on these themes, a semi-structured interview was drafted, planned, pilot-tested and internally validated via mock interviews and focus groups (Ethical Approval from the University of Warwick Ethical BSREC Committee: REGO-2018-2283), and conducted in the south of Benin, across several cities (Calavi, Cotonou, Zinvié), from 22 November to 7 December 2019. Additionally, we obtained informed consent from every participant prior to data collection. The collected answers were then thoroughly analyzed from both a bioethical and sociological viewpoint together with international law experts from the University of Warwick, which led to a discussion on legal transplantation of broad and current interest (presented in the relevant section later in this chapter).

Field Study

During the field study in Benin, 18 people (13 males, five females), aged 25-60 years old, were interviewed in four different hospitals (both public, i.e., CNHU – Centre National Hospitalier Universitaire Hubert K Maga, Hospital d'Abomey-Calavi, and private/religious, such as La Croix Zinvié). Among the interviewees there were nurses, medical doctors, administrative staff, biomedical engineers and technicians (BMETs), members of the Ministry of Health, and scholars considered representatives of the themes that are object of this research. Possible interviewees were selected if they were directly or indirectly involved with MDs for their work and had at least one year of work experience. This means that even if not all the respondents could be defined experts, they can be seen as representatives of people who work daily with such technologies. Out of the 18 interviewees, five were from rural/district hospitals and 13 from the public one. The sample was defined using the reasoned choice and snowball techniques. Reasoned choice provides that the sample is chosen in a way that it represents the studied population as accurately as possible. For this reason, as many representatives of the categories involved with MDs as possible were captured. The sample size was determined by the saturation threshold, i.e., the number of interviewees was stopped when no new codes or themes emerged from the interviews, rather, the same ones started recurring (21, 22).

Overall, we believe that this mixed-background population could be a good representative sample of the Beninese population/healthcare workers.

Results of the semi-structured interviews were extracted by a first manual transcription of the audio recording, followed by a coding via NVIVO, which allowed for the thematic grouping of the answers.

Results

The results are hereby reported by theme.

MDRs : From the interviews, it emerged that the interviewees are aware of the lack of a Beninese MDR, and that the one of references is the French one (i.e., the European Regulation on MDs 745/2017), since Benin is a former French colony. As regards the standards, those of reference depend on the type of the device. To this regard, one of the interviewees reported that, in 2017, Benin ratified the Medicrime Convention[8], issued by the European Union in 2011[9]. The signing of this convention is considered to be the early step towards a new regulation, specific to Benin. Alongside this document, some interviewees also referred to others, concerning MDs in Benin, such as the French Public Health Code relating to biomedical equipment and locations (Sixième partie: Etablissements et services de santé [Articles from L6111-1 to L6441-1]), and the Decree of the Minister of Health of 2013 concerning the management of donations.

Interestingly, according to 94.4% of the interviewees, foreign regulations concerning MDs are not adequate for Benin. In fact, they all claimed to be aware of the need for their contextualization (considering local objective status and subjective perspectives and acceptability), and of the urgency to draft and promote protocols which, while adhering to the reference EU MDR, take into account the particularities of the contexts of applications of such norms.

When respondents were asked if local BMETs share their own concerns over the current status of regulatory frameworks in Benin, they all responded in the affirmative. However, when the question object was shifted to politicians involved in the interview, a criticism was revealed: almost 90% of respondents stated that Beninese politicians have not proposed substantial changes to the existing regulations so far, neither internally nor paying attention to the international community. To this regard, an interviewee explained how the National Strategic Plan relating to MD maintenance, drawn up by the Direction d'Infrastructure d'Equipment et Maintenance (DIEM) of the Beninese Ministry of Public Health, is inspired by European Regulation on MDs 745/2017 and not local needs[10]. One interviewee spoke out their fear hypothesizing that: "*the Beninese authorities do not want their own regulation*". This shows the common belief that the politics prefer to adhere to the existing scenarios, that means existing regulations of the HICs, instead of supporting real local improvement.

Finally, when asked if they perceived the high-resource setting (HRSs)-led drive to identify their own regulations as a form of external imposition, the shared answer was: "*Nemo propheta in patria*", i.e., no man is a prophet in his own land, thus explaining the need to be guided and collaborate in the process of renewal of the regulatory frameworks.

Maintenance: According to 5.6% of respondents, MD maintenance is underestimated. Related to this, the interviews show the gap between the regulatory dimension and the practical/technological dimension: the administrators are not able to fully understand the challenges that are encountered daily in a hospital in the management of MDs (i.e., the lack of spare parts, the inconsistent electricity, instruction manuals in foreign languages, different temperatures and/or application environments very different from those of construction of an MD, and lack of training of technical personnel on new Medical Devices) (17, 19, 20). This scarce level of communication is also present between the administrative/financial staff of a single hospital and the team of engineers and technicians, who are often not informed of newly purchased MDs or novel incoming donations. This causes great confusions, but, above all, shows how the available MDs do not actually meet the needs of patients/users, and adequate management and use protocols are not followed. In fact, most of the interviewed BMETs report that even in the event of a failure of a MD, they are often bypassed for self-modification repairs that take place in an amateurish, and often dangerous, way, or for directly contacting the MD manufacturers. With respect to this, an interviewee acknowledges that in the Beninese Code of Public Health, in the Guide to Good Practices of Biomedical Engineering, it reads that no external technician must intervene on the MDs if it does not go through the internal maintenance service. However, too often the Code is not respected. The reason behind this was given by an interviewee: "*Because people are reluctant to change. The reason is that here they have never seen how the relationship between the maintenance service and the user of the Medical Device should be*". The interviewees reveal how other challenges are a poor consideration of preventive maintenance and quality control of MDs, as it often happens that outdated MDs are not decommissioned. These issues were also noticed during the technical field studies. Only some hospitals presented a preventative maintenance approach, most of them only put into practice corrective maintenance. Moreover, not all the hospitals had a local team of BMETs, some relied on external "on-call" ones (19, 20).

Perception of MDs: This section has been included to evaluate not only the popular perception of MDs per se, but also how this can affect local regulations, effectively responding to the needs of the population, in respect of the contextual culture. Ultimately, the results are positive and underline an overall acceptance of the MDs by the Beninese population. Although there is no real hesitation in allowing one's treatment with MDs, all the interviewees made it clear that the rural population prefers traditional treatments as primary care (e.g., herbal preparations instead of modern drugs). From this, it was clear that there is a lack of healthcare staff in charge of a "cultural mediation", i.e., of making sure that each patient has a clear understanding of how the technical tool (i.e., MD) used for their treatment works. The need for such a role, or for a preventive clarification, which usually falls within the information that the doctor gives the patient during their interaction, becomes crucial in countries where healthcare-patient communication is heavily hindered by illiteracy, language issues (dialects spoken) and, above all, by the peculiar cultural tradition that characterize medical ethics in such places.

Towards a Hermeneutic Heuristic Framework

"*To enjoy benefits of scientific progress and its applications*" is a fundamental human right, which belongs to everyone, even more so if it is linked to health and health technologies, as it is stated by the United Nation's Universal Declaration of Human Rights (art. 27 par. 2) as well as being prioritized by other International Charts (i.e., Article 15 of the International Covenant on Economic, Social and Cultural Rights and the WHO 6 Leadership priorities[11]). In particular, Article 15 of the UNESCO Universal Declaration on Bioethics and Human Rights states: "*Benefits resulting from any scientific research and its applications should be shared with society as a whole and within the international community, in particular with developing countries*", while, as we have seen, the diffusion of health technologies, in particular MDs, is not universal. Most of the world population does not benefit from the use of MDs: less than 15% of the global population accounts for the use of over 75% of the MDs, suggesting inequitable access to healthcare in favor of higher- resource settings (17). The problem is not only related to the access to the healthcare. Some other principles are involved, i.e., Availability, Adequacy, Accessibility, Affordability, and Appropriateness (23), and all of those could impact on the quality, safety, and effectiveness of MDs.

The majority of EU/USA citizens accept and trust MDs and are familiar with their use. Conversely, African traditional healthcare is less prone to the use of devices. Moreover, the majority of MDs are designed by and for EU, USA, and Japan, who account for almost 80% of the global MD market (24). Europe, USA, and Japan have homogeneous medical knowledge, clear standards, harmonized regulations, and trade-agreements allowing free circulation of MDs among hospitals/countries, while maintaining the same level of safety and efficacy. Consequently, designers and regulators take HRSs hospital infrastructures (e.g., the ones in EU and USA) for granted, completely ignoring the challenges of LRSs.

Some authors have argued that the European Regulation on MDs 745/2017, USA FDA, and Japanese Pharmaceutical Affairs Law are not evidence-based (25), and have historically been written for economic and commercial reasons, and not only to ensure patient safety (26, 27). Conversely, the African MD manufacturing is not prominent, and African regulations are not homogeneous (28), and often not sufficiently adhered to nor monitored. Unfortunately, USA and EU regulations, despite the allegedly universal criteria, do not consider LRSs local cultures and specific conditions (25). This evidently intensifies the disparities and inequalities among populations, injuring human dignity and rights. This reinforces the idea of a western healthcare paternalism that, as an intellectual neo-colonialism, wants to be an imposition of principles and values that are allegedly universal, but as a result are contextualized and not aware of objective (climatic and hygienic conditions) and subjective-cultural (the self-representation and identity of different populations).

It could be hypothesized that MDRs rely on some political and economic choices, inspired by Adam Smith's libertarianism, i.e., the idea of free trade (bound to the idea of liberty, of the "*laissez-faire*"), which is different from Locke's liberalism (related to the idea of freedom and to the defense of human rights) (29). Accordingly, the contained standards and norms will surely have some ethical reasons, but also economical ones, i.e., the idea to easily sell in the common market. This caused competitiveness among "rich countries", which led to the definition of minimum requirements and criteria impossible to be met by LMICs. As a consequence, most of the MDs designed by HIC are bound to a short-term survival in LRSs[12], and MD effectiveness and safety is no longer guaranteed, thus, for a part of the world, principles and rights are not guaranteed either. In this regard, the most compromised principle is that of appropriateness, as MD design is often contextualized and this compromises the universalism of norms and standards. On the other side, this attitude results in a defensive closure of the populations in LRSs towards innovative medical healthcare technologies, which are perceived as extraneous.

Regulations should also take into account the fact that different countries have a different historical context, cultures, traditions, perception of healthcare, population priorities, individual perspectives and needs, and ethics. However, this should not lead to a relativistic approach that does not allow unitary perspectives and it can be dangerous, especially for the rights of the most vulnerable people (30). Conversely, the correct approach should be universalistic, relying on more than one perspective and several principles (other than just economical) and rights that must be considered in their specificity. The aim is, in fact, twofold, i.e., empowering LRSs, and changing perspective towards a reframing of democracy. It is only by expanding the concept of universality, considering simultaneously the particularity of different contexts, with their subjective and objective local aspects, that a full and cosmopolitical citizenship of rights and social equality will be reached.

To address this aim, the authors of this article are working on the definition of a "hermeneutic heuristic framework". Hermeneutic means interpretative, i.e., able to observe things from a peculiar perspective, that of the interculturalism, and not of the multiculturalism. Heuristic refers to the underlying methodology: in fact, this framework will be based on the empiricism of inductive reasoning or the "by design" approach that is particularly functional as it allows us to design new work proposals starting from specifics. The envisioned framework is interdisciplinary: BME, philosophy, law, ethics are combined to build a strategy of negotiation among disciplines, and among particularism and universalism.

The case study of Benin helped the authors highlight the importance of the particularistic and specific aspects for each context. Benin was exemplary for the deeply rooted traditional religion and culture that has a direct impact on the perception and acceptance of healthcare technologies. Benin is also exemplary because it has no regulation of its own and tries to be aligned with the European one, as it was a former French colony. For all these reasons, Benin is the elective country to represent the main challenges that the authors had identified concerning MDR. This allowed the authors to analyze them with a contextualized approach and propose such framework to overcome the limits of the current situation.

Far from extreme particularism and universalism, and following intercultural mediation, the framework will rely on the moderate or contextualized universalism, as formulated by Martha Nussbaum and Amartya Sen, who theorized a philosophical and economical model of universalism of rights that takes into account individual "capabilities" (i.e., capability approach). This approach, as suggested by the name, focuses on human capabilities, i.e., what people are actually capable of doing and being, and fights discrimination and unequal treatment, being a fully universal approach (31–33).

According to Nussbaum, this approach can be used to generate political principles by the joint effort of economists, policymakers, political scientists, and so on, supported by multinational corporations, global economic policies, agencies agreements, international bodies, and non-governmental organizations. In fact, she strongly believes that each country can pragmatically leverage its capabilities, keeping in mind the SDGs, for a fairer world.

Accordingly, our framework will guide:

Biomedical engineers and MD designers from EU and Africa in considering key gaps between LRSs and HRSs affecting MDs' safety and effectiveness.

International regulators fostering the harmonization across LMICs and HICs, which will be possible only if there will be a cooperation among Africa- and EU-based experts.

Policymakers in defining more universal regulations that are not only economy-driven, but prioritize the safeguarding of human rights and dignity.

In fact, the referral to human rights must not be misunderstood as a new type of "transplant", rather as a tensional reference, an aim to a "universal" that must be taken into consideration in a "situational" way, i.e., by negotiating with the local contexts and their capabilities (34).

This framework (see Figure 3.3) is also crucial to improve communication between scientists and politicians in the public debate. Politicians, in fact, should rely on a more solid scientific culture, as scientists should on a more solid political culture, growing awareness of their social role. In this way, the framework, with its multidisciplinary ramifications, i.e., political, ethical, sociological, and technological, could help the increasing of the universality of MDRs, broadening the access to the full range of human rights and, at the same time, with the help of the biomedical engineer and ethics by design and the frugal regulation approach, address the particularism of each context, respecting and protecting the most vulnerable. The political justice theorist Nancy Fraser argued about the possibility of "reframing justice" (35). Along the same lines, but starting from a broader multidisciplinary and intercultural perspective, the authors of this hermeneutic heuristic framework think that following this theoretical-practical approach will allow to rethink and, therefore, to redefine and reframe the concept of democracy, including in a transformative way all the cultural specificities and the different

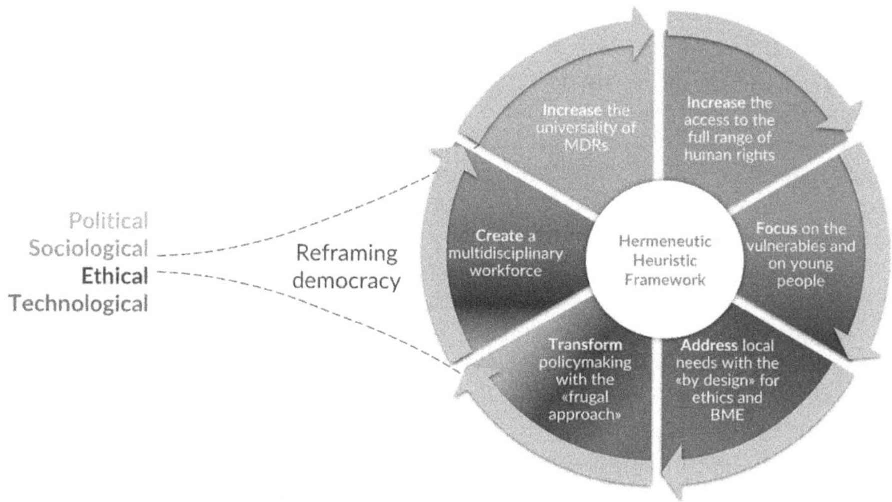

FIGURE 3.3
The objectives of the hermeneutic heuristic framework, divided by category, i.e., political, sociological, ethical, and technological.

social/professional categories mentioned above (legal, political, techno-scientific, etc.) at national and global level.

Discussions: Legal Transplantation or Frugal Regulation?

The previous section laid the basis for our framework. This section will present a put-into-action plan that will follow a heuristic methodology. The first results (engineering related) of this extended study have already been published (17, 19). Specifically, in (17) a "by design" approach was followed to guide the design of MDs resilient to LRSs. On the same basis, the authors believe that the "by design" approach is also applicable to the sociological and bioethical perspectives. "Ethics by design" is an existing discipline that finds ethical questions based on the characteristics (e.g., tools and algorithms) of novel technologies (36, 37). However, the interdisciplinary nature of this study led us to introduce "bioethics by design", i.e., an approach that refers to the concept of life[13], and, therefore, focuses the design on the needs of the human beings and the multiple questions that emerge from the use of biotechnologies (38, 39). Although this approach is still being discussed (40), it is well suited to the present study, because it brings us closer to the problem of technologies (MDs), and their regulations, that should be built following a "bioethics by design" approach, i.e., built on local needs.

But how do laws and regulations become universal and particular at the same time? Shifting our focus on a legal and policymaking level, we can make a historical reconstruction. With respect to the MDs, it should be clarified that the European Regulatory Frameworks from the seventies to the nineties experienced a transition from a subjective and prescriptive approach to a more global, objective one that gradually made producers responsible. The change is witnessed by the fact that also lexically we have recently passed from "directives" to "regulations", more rigid and uniform than the former. Furthermore, the new regulations are a comprehensive legal instrument, which no longer requires national transposition and is directly applicable and mandatory in each Member State (3).

This is a challenge for Africa, particularly for a countries like Benin, where there are no local regulations, as:

There is no drive to draw up and communicate at an international level their own local needs.

They often refer to the criteria indicated by the manufacturer, which do not always meet the local conditions.

De facto inapplicable regulations are kept in force.

It should be noted that the trend of displacing and imposing a body of norms and laws from its country of origin to another external country is a phenomenon better known as "legal transplantation" (41).

Legal transplantation theory provides the contexts for cultural difference, the feasibility of legal borrowing, and law and development for the regulatory efficient analysis where regulations/standards are borrowed. The terminology of legal transplantation, also referred to as legal borrowing and legal transfer, was first provided by Alan Watson (1974) (42). Watson identified transplants as "*the moving of a rule or a system of law from one country to another*". Meanwhile, Langer (43) labels transplants as the phenomenon of circulating legal ideas and practices. Geoffrey Samuel (44) terms legal transplants as an approach in comparative law. In understanding the implementation of foreign/international regulations/standards, such as those concerning MDs, therefore, it is useful to examine the legal transplants theory. This will shed light on whether the universal regulations/standards are valid.

The debate on legal transplants is confined to questions as to whether transplantation has been successful. Watson argues that the process of legal transplants in a foreign country is quite simplistic in nature and can be achieved with minimum scholarly expertise (42). Interestingly, Watson claims that transplanted laws have no links with any socio-cultural influences within a country and that the laws in no way mirror society at large (45). Conversely, Legrand (1997) argues that legal transplants are impossible and can never be successful (41). He points out that law is directly derived from a socio-cultural framework, and hence, legal transplantation could vary between cultures and societies. For Legrand the law is heavily influenced by a socio-cultural context, and any interpretations of the law can only be culturally determined, and that a "*rule does not have any empirical existence that can be significantly detached from the world of meanings that characterizes a legal culture*" (46). Moreover, Teubner argues that the theory put forth by Watson on the distinction between law and society is in direct conflict with social and cultural theories. Several other scholars have had contrasting opinions in this regard (47–51). Overall, the notion of legal transplants in applying regulations to another country is not valid unless other factors such as culture, social differences, and the country of origin are also considered by regulators.

Regarding the question concerning the success of legal transplantation, there is no standard of measurement on successful receptions. Although it is difficult to formulate a measure for success and to determine the success or failure of legal transplants, one factor in successful legal transplantation is whether the country is receptive to the maintenance and support of transplanted laws (52). The process of legal transplantation could be controversial in terms of the responsiveness to legal traditions under new conditions as they need to be "context-specific". The transplanted laws can only be effective as long as they remain "local" (53).

If legal transplants are to be effective in their design, their implementation has to be considerate of local cultures, religions, socio-economic-political factors, as well as geographic differences (49). As the results vary significantly on a case by case basis regarding which laws can be used for transplants, Kahn-Freund points out that transplants are possible only if the laws can be alienated from their origins (49). He emphasizes that certain laws are highly rooted in their origins and, thus, cannot be transplanted to other cultures. Moreover, the success of legal borrowing can be determined simply in terms of whether transplanted laws are implemented according to the intentions of the legislators. Therefore, the success of a legal transplant depends on the ability of the host country to adapt foreign regulations and standards to its local conditions and that the transplanted law can serve the objectives of the legislation.

Overall, it is critical to observe and investigate the impact of legal transplantation. The effects of transplants could have unintended consequences on an economy.

In this regard, there is a growing literature on the failure of legal transplants which focuses on why transplanted laws fail in emerging economies (54–56). One reason for this is that the borrowing of regulations and standards that originate from developed countries may be contradictory to their own systems. For example, they may be due to the political pressures and good will to enforce international standards on all countries. Cranston (57) emphasizes his concerns on blind transplants of the standards, codes, and principles of new international financial law or, as he terms it, "*neo-colonial domination*". This is important because from another aspect, emerging countries are forced politically to apply the standards of new regulations without consent. Legrand finds that legal transplants should theoretically be aimed at the benefit of society and not for economic advances. Thus, laws favoring developed countries are transplanted into these developing nations to attract foreign direct investments, which boost the economy. Legrand's contention is that legal transplants must consider local cultural attributes by being aligned to the globalization process, and foreign laws are now gaining more acceptance at the local level (58). As globalism creates new cultural and social norms, developed countries often tend to take advantage of these norms to create powerful political and economic institutions at the international level (58).

While the debate on legal transplant has found both critics and backers, experts such as Legrand find that legal transplants are an impossible feature in the legal system as they need to cater to the society they are being transplanted into, which may be totally alien to the society the law is being transplanted from. However, other experts such as Watson clearly believe that legal transplants are possible and are not necessarily affected by any impact that socio-cultural factors may play in legal transplants. Legrand argues strongly that legal transplants can cause adverse effects on society and result in undesirable outcomes and, therefore, they must be rejected. However, the available literature on the subject finds that despite its many challenges, there is a role for legal transplants in reforms of the law. Nevertheless, the many challenges facing legal transplants, especially due to globalization, must first be addressed for transplants to become successful.

The fact that the transplant itself refers to the notion of colonialism and the contextual need for a reasoning that still points to the universal (human rights, global health) without being generic nor oblivious to particular contexts, invites us to refer to Herrick et al. and Tuck et al. (59, 60), who consider decolonization, in this case for global health, as something serious and not metaphorical

In fact, while these debates on legal transplantation continue, a different ethical-legal tool is proposed for our study, namely that of "frugal" regulations. The definition "frugal" comes from "frugal innovations", that is the development of appropriate, adaptable, affordable, and accessible goods, services, and solutions especially for the non-affluent customers, in particular in the context of emerging markets (61, 62). According to Weyrauch and Herstatt "frugal innovation" is defined by three criteria: substantial cost reduction, concentration on core functionalities, and optimized performance level (63). Precisely for this reason, we think this concept of frugality is applicable to regulations so that they are sustainable, functional, and optimized according to subjective and objective local needs. Furthermore, when referring to "frugal regulations" we mean a provisional structure, designed according to the design of specific needs, and flexible or fluid. This "frugal regulation" is inspired by a learn-by-doing approach, or by the progressive modification and improvement that takes into account the specific context in constant change.

"Frugal regulations" represents the last building block of the proposed framework (see Figure 3.4) because it envisions a contextualization rather than a self-regulation for each country. This means a negotiation and a by design adaptation of the commonly stated principles and norms within each culture and local setting, without falling into relativism. Figure 3.4, in particular, shows how the research question emerged and the steps taken to answer it. The multidisciplinary approach, which allowed the use of different methodologies which shared a bottom up approach. This is particularly evident in the "by design approach", which is typical of the domains of ethics and engineering. The research team, therefore, followed an approach that allows the consideration of local factors, objective (i.e., MD location and settings, existing ethical principles, and regulation in force, if any) and subjective ones (i.e., perception of

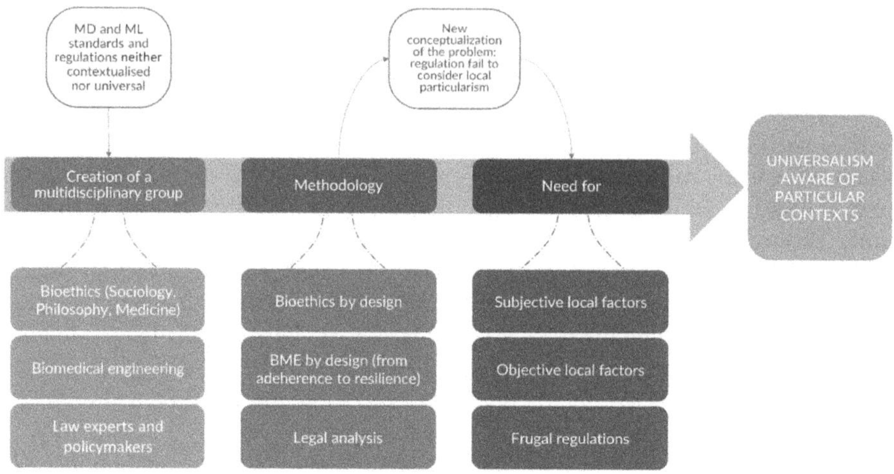

FIGURE 3.4
The process behind the creation of the hermeneutic heuristic framework, including methodologies and objectives.

healthcare technologies, local traditions of care, and traditional culture). This can be considered as a "frugal regulation" approach. With this last piece of the puzzle, it seems possible to propose a new universalism aware of the different specific contexts and, therefore, a flexible regulation that can explain different cultural and traditional contexts without losing its universality and falling into relativism.

The validation of this framework validation is still ongoing, for example:

At the scientific community level, authors have already brought the subject up with the European Alliance of Medical and Biological Engineering and Science (EAMBES) and built a working group that is supporting the WHO on the definition of a nomenclature for MDs that is universal but aware of particular context.

At the MD design level, D.P. is working on several use cases, some of which have been published recently (64, 65), and he is also spreading the design philosophy presented in (17) at several summer schools[14]. This design philosophy is aligned with the overarching concepts presented in this paper and its validation will, therefore, be a crucial step for the validation of this framework.

The International Federation of Medical and Biological Engineering (IFMBE), supported by L.P., launched the first IFMBE Africa biomedical engineering working group. The latter, supported by the WHO and African Union[15], reunites for the first time 20 African National Scientific societies of BMETs, offering an unprecedented opportunity to dialogue with African experts on MDs and medical locations.

In 2018, in Ethiopia, L.P. witnessed the birth of the African Community of BME, which aims to bring the African voice to and make it heard at EU level.

In 2018, L.P., during the European Parliament Interest Group on Biomedical Engineering, organized a meeting to identify possible political actions aiming to reduce the gaps among Sub-Saharan and European countries, promoting the harmonization of MDs and medical locations regulations. As a result, a task force of the IFMBE wrote down the Health Technology Assessment (HTA) Guidelines for Medical Devices.

Networking and collaborations with African scholars in Uganda, Benin, South Africa, Ethiopia, and Mozambique.

Collaboration with WHO experts (such as Adriana Velasquez and Dr. Sheick Oumar Coulibaly).

Collaborations with international scientific societies for Health Technology Assessment (Health Technology Assessment International [IHTA] and International Society for Pharmacoeconomics and Outcomes Research [ISPOR]), scientific societies of MD designers (IFMBE, EAMBES, and the International Union for Physical and Engineering Sciences in Medicine [IUPESM]) and manufacturers, such as Med Tech.

Writing scientific publications on the matter and participation in national and international congresses[16].

Capacity building in Benin, holding lectures at the University of Abomey-Calavi, raising the students' awareness of the needs.

Conclusion

Our field study showed how the local population, and the experts on the subject, are perfectly aware of the MDR problem, and that they require guidance to get out of the impasse, that is, for regulations that are more responsive to local objective and subjective needs. The first results revealed the importance of re-evaluating the role of maintenance, gendered norms, local medical ethics, and prompting politicians to focus more on the local needs rather than on international issues.

The authors followed a bottom-up methodology: starting from "ethics by design", they proposed a "bioethics by design" that is capable of holding together the interdisciplinary gaze of bioethics and the methodology of design to build an interpretative framework that is effectively able to approach specific contexts and, at the same time, to strive for the universal dimension. Our hermeneutic heuristic framework, in fact, by interpreting the contextual reality in its specific particularity with quantitative and qualitative methods, addresses the political, regulatory, technical, and design side. This framework allows LRSs inhabitants themselves to be able to evaluate their own needs, bring them to collective attention, and propose new endogenous strategies for action: in fact, as it suggests to the designers a more attentive approach to the *resilience* (to LRSs) of MDs rather than their *adherence* (to the currently existing standards), at the same time, it suggests to policy makers the use of "frugal regulations", i.e., rules that are never uniquely defined and always sensitive to change, and to the scientific and cultural progress of a country. This article tried to take stock of and summarize the pieces of a challenging study of which this is but a part.

Limitations

One might argue that one of the limitations of this study is the limited sample size, which may not be representative of the whole Beninese populations/healthcare workers. However, both the fact that our sample includes people with different cultural and socio-economic backgrounds (different level of experience, education, and positions, etc.) and our use of the aforementioned saturation threshold to capture all the possible themes, make this limitation negligible. Nonetheless, further comparative ethnography studies could focus on other populations of Sub-Saharan Africa to see whether similar beliefs and perceptions are shared within wider communities.

List of abbreviations

MDs: Medical devices
BME: Biomedical engineering
BMETs: Biomedical engineers and technicians
LRSs: Low-resource settings
HRSs: High-resource settings
LMICs: Low- and middle-income countries
HICs: High-income countries
SDG: Sustainable Development Goal
MDRs: Medical device regulations
WHO: World Health Organization
USA: United States of America
PIP: Poly Implant Prothèse
FDA: Food and Drug Administration
NRAs: National Regulatory Agency
UEMOA: West African Economic and Monetary Union

GMB: Départment du Génie Biomédical- Benin
EPAC: Ecole Polytechnique d'Abomey-Calavi
LAMA: Laboratoire de Anthropologie Médicale Appliqué – University of Abomey-Calavi
ABSPIE: Applied Biomedical Signal Processing and Intelligent e-Health Lab – University of Warwick
WICID: Warwick Interdisciplinary Research Centre for International Development – University of Warwick
ICEHTMC III: Third International Clinical Engineering and Health Technology Management Congress
CNHU: Centre National Hospitalier Universitaire Hubert K Maga
DIEM: Direction d'Infrastructure d'Equipment et Maintenance
IHTA: Health Technology Assessment International
ISPOR: International Society for Pharmacoeconomics and Outcomes Research
IFMBE: International Federation of Medical and Biological Engineering
IUPESM: International Union for Physical and Engineering Sciences in Medicine
EAMBES: European Alliance for Medical and Biological Engineering and Sciences

Declarations

Ethics approval and consent to participate: All methods were performed in accordance with the relevant guidelines and regulations. This study received full ethics approval from the University of Warwick Ethical BSREC Committee: REGO-2018-2283. Informed consent was collected from the participants prior the start of the interview.

Notes

1 A. Maccaro, D. Piaggio, S. Leesurakarn, S. Sekalala, S. M. Rai, L. Pecchia, On the universality of Medical Device Regulations: the case of Benin, in «BMC Health Services Research», (2022) 22:1031, https://doi.org/10.1186/s12913-022-08396-2 (urly.it/3pryx)
2 www.imdrf.org/
3 Benin, Burkina Faso, Ivory Coast, Guinea-Bissau, Mali, Niger, Senegal, and Togo.

4 www.uemoa.int/en/renforcement-du-dispositif-reglementaire-des-produits-et-technologies-de-sante-dans-l-espace-uemoa
5 Multiculturalism, in defending cultural diversity and protecting the rights linked to the identity of each culture, ends up understanding cultures as monolithic identities, making the cultural exchange difficult; Interculturalism assumes that every culture has intrinsic historicity, which opens it up to endogenous modifications and exchanges with others, thus not perverting the sense of the cultural identity of each (16. Levrau F, Loobuyck P. Introduction: mapping the multiculturalism-interculturalism debate. SpringerOpen; 2018. p. 1-13.
6 Medical devices in LMICs: global gaps, trend and opportunities (L.P.); A device for tracking the photo-pupillary reflex via smartphone in LMICs (D.P.); Medical devices in Africa: ethical aspects and universality of norms (A.M.); Development of IOT-based patient health monitoring and management system (Busola Oronti)
7 How can countries develop Policy and help implement Solutions (Adriana Velazquez, Jitendar Sharma, Tobey Clark, Paolo Lago, Nicolas Pallikarakis, Renato Garcia, Roberto Ayala); The need for integration among BME and MPs, the WHO prospective (Adriana Velazquez Berumen)
8 www.coe.int/en/web/medicrime/signatures-/-ratifications
9 www.edqm.eu/en/Medicrime-Convention-Background
10 www.pnt-benin.org/wp-content/uploads/2019/02/Benin_PSN_TB_2015-2019.pdf
11 www.who.int/about/resources_planning/WHO_GPW12_leadership_priorities.pdf
12 Resolution WHA60.29 about Health Technologies in Sixtieth World Health Assembly, 2007
13 From the Greek βίίς, which means "life".
14 African Biomedical Engineering Consortium Design school (Kampala, Uganda, 2019), Italian National Bioengineering Group (BNG) Annual Bioengineering School (Brixen, Italy, 2022).
15 IFMBE. Kick-off of the First African scientific society of Medical and Biological Engineering 2018 [Available from: http://htad.ifmbe.org/kick-off-of-the-first-african-scientific-society-of-medical-and-biological-engineering/.
16 IUPESM World Congress (Prague, 2018), IV WHO Global Forum on Medical Devices (Visakhapatnam, 2018), ICEHTMC (Rome, 2019), CMBEBIH (Banja Luka, 2019).

References

(1) WHO. Regulatory system strengthening for medical products. *Proceedings of the 67th World Health Assembly*. 2014;28.

(2) Helyar V, Burke C, McWilliams S. The ruptured PIP breast implant. *Clinical Radiology*. 2013;68(8):845–850.

(3) Pecchia L, Piaggio D, Castaldo R, Radice L, Pallikarakis N. Medical device regulation and assessment: New challenges for biomedical engineers. Pàtron Editore. *XXXVIIIa Scuola di Bressanone*;2019, pp. 177–192.

(4) FDA. A history of medical device regulation and oversight in the United States. Available from: www.fda.gov/medical-devices/overview-device-regulation/history-medical-device-regulation-oversight-united-states

(5) Directive C. 90/385/EEC of 20 June 1990 concerning active implantable medical devices. *Official Journal of the European Communities*. 1990.
(6) Directive C. 93/42/EEC of 14 June 1993 concerning medical devices. 1993.
(7) Herman P, Horowitz J, Torsekar M. *Competitive conditions affecting us exports of medical technology to key emerging markets*: US International Trade Commission; 2018.
(8) Altenstetter C. Medical device regulation in the European Union, Japan and the United States. Commonalities, differences and challenges. *Innovation: The European Journal of Social Science Research*. 2012;25(4):362–388.
(9) Statista. Size of the medical device market in Africa in 2017 and a forecast for 2023. Available from: www.statista.com/statistics/890876/medical-device-market-size-africa/
(10) WHO. *Global atlas of medical devices*: World Health Organization; 2017.
(11) Kedwani M, Schröttner J, Baumgartner C. Analysis of regulatory requirements of medical devices and in-vitro diagnostics worldwide for the development of an efficient procedure of registration for manufacturers of medical products. *Current Directions in Biomedical Engineering*. 2019;5(1):609–612.
(12) Gobbers D, Pichard É. L'organisation du système de santé en Afrique de l'Ouest. *Actualité et dossier en santé publique*. 2000 Mar:35–40.
(13) Palgo DH. *L'harmonisation du droit pharmaceutique en Afrique de l'Ouest: le cas de l'UEMOA*: Bourgogne Franche-Comté; 2018.
(14) Maccaro A, Piaggio D, Dodaro CA, Pecchia L. Biomedical engineering and ethics: Reflections on medical devices and PPE during the first wave of COVID-19. *BMC Medical Ethics*. 2021;22(1):1–7.
(15) Levrau F, Loobuyck P. *Introduction: Mapping the multiculturalism-interculturalism debate*: SpringerOpen; 2018. 1–13.
(16) Polisena J, Castaldo R, Ciani O, Federici C, Borsci S, Ritrovato M, Clark D, Pecchia L. Health technology assessment methods guidelines for medical devices: How can we address the gaps? The international federation of medical and biological engineering perspective. 2018 Jan;34(3):276–89.
(17) Piaggio D, Castaldo R, Cinelli M, Cinelli S, Maccaro A, Pecchia L. A framework for designing medical devices resilient to low-resource settings. *Globalization and Health*. 2021;17(1):1–13.
(18) Pope C, Mays NJB. Qualitative research: Reaching the parts other methods cannot reach: An introduction to qualitative methods in health and health services research. *BMJ*. 1995;311(6996):42–45. https://doi.org/10.1136/bmj.311.6996.42
(19) Di Pietro L, Piaggio D, Oronti I, Maccaro A, Houessouvo RC, Medenou D, et al. A framework for assessing healthcare facilities in low-resource settings: Field studies in Benin and Uganda. *Journal of Medical and Biological Engineering*. 2020;40(4):526–534.
(20) Piaggio D, Medenou D, Houessouvo RC, Pecchia L, editors. Donation of medical devices in low-income countries: Preliminary results from field studies. In CMBEBIH 2019: Proceedings of the International Conference on Medical and Biological Engineering, 16-18 May 2019, Banja Luka, Bosnia and Herzegovina 2020 (pp. 423–427). Springer International Publishing.
(21) Saunders B, Sim J, Kingstone T, Baker S, Waterfield J, Bartlam B, et al. Saturation in qualitative research: Exploring its conceptualization and operationalization. *Quality & Quantity*. 2018;52(4):1893–907.

(22) Baker S, Edwards R. How many qualitative interviews is enough? Expert voices and early career reflections on sampling and cases in qualitative research. National centre for research methods review paper. 2017.
(23) McLaughlin CG, Wyszewianski L. Access to care: Remembering old lessons. *Health Services Research*. 2002;37(6):1441–1443.
(24) MedTech Europe. The European Medical Technology Industry in figures 2021. MedTech Europe Brochure. 2021.
(25) Fahlgren B. *Access to effective medical technology in developing countries – What role for WHO*: WHO; 2004.
(26) Richards-Kortum R, Oden M. Devices for low-resource health care. *Science*. 2013;342(6162):1055–1057.
(27) Kouris K, Abdel-Dayem H. Transfer of medical technology from a developed to a developing country. *Journal of Biomedical Engineering*. 1988;10(4):326–330.
(28) Biehl J. *When people come first: Beyond technical and theoretical quick-fixes in global health*. Global Political Ecology: Routledge; 2010. pp. 114–144.
(29) Barry N. *On classical liberalism and libertarianism*: Springer; 1987.
(30) Renteln AD. *International human rights: Universalism versus relativism*: Quid Pro Books; 2013.
(31) Nussbaum M, Sen A. *The quality of life*: Clarendon Press; 1993.
(32) Nussbaum MC. *Frontiers of justice: Disability, nationality, species membership*: Belknap Press; 2006.
(33) Nussbaum MC. *Women and human development: The capabilities approach*: Cambridge University Press; 2001.
(34) Sen A. Human rights and capabilities. *Journal of Human Development*. 2005;6(2):151–166.
(35) Fraser N. *Reframing justice*: Uitgeverij Van Gorcum; 2005.
(36) Shilton K. Values levers: Building ethics into design. *Science, Technology, & Human Values*. 2013;38(3):374–397.
(37) Dignum V, Baldoni M, Baroglio C, Caon M, Chatila R, Dennis L, Génova G, Haim G, Ließ MS, Lopez-Sanchez M, Micalizio R. editors. Ethics by design: Necessity or curse? In Proceedings of the 2018 AAAI/ACM Conference on AI, Ethics, and Society 2018 Dec 27 (pp. 60–66).
(38) Pavarini G, McMillan R, Robinson A, Singh I. Design bioethics: A theoretical framework and argument for innovation in bioethics research. *The American Journal of Bioethics*. 2021;21(6):37–50.
(39) Brall C, Schröder-Bäck P, Maeckelberghe E. Ethical aspects of digital health from a justice point of view. *European Journal of Public Health*. 2019;29(Supplement_3):18–22.
(40) Martschenko D, Martinez-Martin N. What about ethics in design bioethics? *The American Journal of Bioethics*. 2021;21(6):61–63.
(41) Legrand P. The impossibility of 'legal transplants'. *Maastricht Journal of European and Comparative Law*. 1997;4(2):111–124.
(42) Watson A. *Legal transplants: An approach to comparative law*: University of Georgia Press; 1993.
(43) Langer M. From legal transplants to legal translations: The globalization of plea bargaining and the Americanization thesis in criminal procedure. *Harvard International Law Journal*. 2004;45:1.

(44) Samuel G. All that heaven allows: Are transnational codes a 'scientific truth'or are they just a form of elegant 'pastiche'? 2012: 165–191.
(45) Watson A. *Society and legal change*. 2nd ed.: Temple University Press; 2010.
(46) Legrand P. John Henry Merryman and comparative legal studies: A dialogue. *The American Journal of Comparative Law*. 1999;47(1):3–66.
(47) Ewald W. Comparative jurisprudence (II): The logic of legal transplants. *The American Journal of Comparative Law*. 1995;43(4):489–510.
(48) Seidman A, Seidman RB, Waelde TW. *Making development work: Legislative reform for institutional transformation and good governance*: Springer; 1999.
(49) Kahn-Freund O. On uses and misuses of comparative law. *Modern Law Review*. 1974;37:1.
(50) Nelken D. The meaning of success in transnational legal transfers. *Windsor Year Book of Access to Justice*. 2001;19:349.
(51) La Porta R, Lopez-de-Silanes F, Shleifer A. The economic consequences of legal origins. *Journal of Economic Literature*. 2008;46(2):285–332.
(52) Ma T-C. Legal transplant, legal origin, and antitrust effectiveness. *Journal of Competition Law and Economics*. 2013;9(1):65–88.
(53) Michaels R. "*One size can fit all" – Some heretical thoughts on the mass production of legal transplants*. Order from Transfer: Edward Elgar Publishing; 2013.
(54) Pistor K, Raiser M, Gelfer S. Law and finance in transition economies. *Economics of Transition*. 2000;8(2):325–368.
(55) Nichols PM. Viability of transplanted law: Kazakhstani reception of a transplanted foreign investment code. *University of Pennsylvania Journal of International Law*. 1997;18(4):1235.
(56) Xanthaki H. Legal transplants in legislation: Defusing the trap. *International & Comparative Law Quarterly*. 2008;57(3):659–673.
(57) Cranston R. A theory for international financial law? *Student Bar Review*. 2006;18(1):6–9.
(58) Gerber DJ. Globalization and legal knowledge: Implications for comparative law. *Tulane Law Review* 2000;75:949.
(59) Herrick C, Bell K. Epidemic confusions: On irony and decolonisation in global health. *Global Public Health*. 2022 Aug 3;17(8):1467–78.
(60) Tuck E, Yang KW. Decolonization is not a metaphor. *Decolonization: Indigeneity, Education & Society*. 2012;1(1).
(61) Basu RR, Banerjee PM, Sweeny EG. Frugal innovation: core competencies to address global sustainability. Journal of Management for Global sustainability. 2013;1(2):5.
(62) Soni P, Krishnan RT. Frugal innovation: Aligning theory, practice, and public policy. *Journal of Indian Business Research*. 2014 May 14;6(1):29–47.
(63) Weyrauch T, Herstatt C. What is frugal innovation? Three defining criteria. *Journal of Frugal Innovation*. 2017;2(1):1–17.
(64) Piaggio D, Namm G, Melillo P, Simonelli F, Iadanza E, Pecchia L. Pupillometry via smartphone for low-resource settings. *Biocybernetics and Biomedical Engineering*. 2021 Jul 1;41(3):891–902.
(65) Piaggio D, Andellini M, Taher M, Pecchia L, editors. A vest for treating jaundice in low-resource settings. In *2021 IEEE International Workshop on Metrology for Industry 40 & IoT (MetroInd40 & IoT)* 2021 Jun 7 (pp. 122–127). IEEE.

Chapter 4

Building a Stronger Europe: The EAMBES' Role in Healthcare Resilience and Sustainability

Background

The last century witnessed an unprecedented revolution in the healthcare sector due to the introduction of engineering into the health field: thus, a specific discipline gradually defined itself, capable of integrating the highly technological skills of engineering with the domain of medicine for preventive, diagnostic, interventional, and healing purposes. Biomedical engineering (BME) thus becomes part of healthcare, providing tools that are redefining the contemporary semantics of care through solutions that, by definition, exceed predictability.

If we were to compare two operating rooms, one from the 1800s and one contemporary, we might notice that the healthcare team remains the same (surgeon, anesthetist, nurses, and surgical technologists), what changes is the environment, which is indispensably characterized by the presence of biomedical technologies that comply with international standards safeguarding patient safety and the success of the procedure[1].

It is necessary to clarify that the term "health technology" refers to "any intervention that can be used for the safe and effective prevention, diagnosis, treatment,

DOI: 10.1201/9781003436348-13

and rehabilitation of diseases and ailments. This category includes: drugs (from an engineering perspective, for example, regarding administration); medical devices, as defined by European directives; procedures (such as surgical techniques, acupuncture, medical counseling); secondary care (such as hospitals, clinics); health programs (screening programs, public health initiatives, etc.)"[2]. The introduction of such technologies in the field of medicine has led to an unparalleled healthcare revolution to the extent that it can be argued that "biomedical engineering is not simply a branch of modern medicine. Modern medicine achieves significant advances primarily through the use of biomedical engineering products"[3].

Currently, there is a "dizzying growth of the global medical device market. A possible indicator of this growth is the number of patent applications filed each year. In fact, while pharmaceutical research is growing very rapidly with about 6,000 new patent applications every year from 2006 to 2016, the number of patent applications for medical technologies increased from 8,000 per year in 2006 to over 12,000 new patent applications in 2016 (Europe, 2018). Moreover, the time to market for a medical device is becoming increasingly shorter. These two factors suggest that the world is about to face a 'tsunami' of medical devices, which will further revolutionize medicine and the way we approach health and wellness management"[4].

A consequence of this "tsunami of medical devices"[5] is that "this has created an unprecedented global need for well-trained biomedical engineers who can help healthcare systems assess devices"[6]. Indeed, there is growing attention around the profession of the biomedical engineer, who "has a direct impact on the quality of life, safety, and cost-effectiveness of healthcare services, thanks to their commitment in the field of research, innovation, evaluation, procurement, and management of the main products of biomedical engineering, such as biotechnologies and medical technologies (MedTech) like medical devices, personal protective equipment, and digital health"[7].

The scope of activities for biomedical engineers is very significant and particularly expanded between the 1950s and 1960s to include multiple activities such as the detection, measurement, and monitoring of physiological signals (e.g., biosensors and biomedical instrumentation), diagnostic interpretation through bioelectric signal processing techniques, and much more. Indeed, BME involves the application of knowledge, skills, and approaches from all engineering disciplines (electrical, mechanical, and chemical) to health problems. The role of biomedical engineers is difficult to fully explain[8]. The multiplicity of variants in the definition of this type of engineering causes quite a bit of confusion: bioengineering, biomedical engineering, biological and clinical (or medical) engineering, all present in what represents a classic text of the discipline, namely the volume "Bioengineering Education Directory" by Pacela[9].

However, biomedical engineers have needed to join specific working groups and international associations to be recognized as a strong, autonomous profile: indispensable both for the design, development, and assessment of new technologies and for their regulation.

The EAMBES Aim

With the aim of establishing an international unit of experts in biomedical technologies, the European Alliance of Medical and Biological Engineering and Science (EAMBES) was founded in 2003 and incorporated under the Belgian law, because its headquartered is in Brussels, Belgium[10]. Important is the name *alliance*, instead of association, federation, etc., because it is a non-profit international organization, whose main objective is to improve health, wealth, and well-being of European citizens by promoting the application of medical and biological engineering and science (MBES). The EAMBES emphasizes the broader term "Medical and Biological Engineering and Science" instead of just "Biomedical Engineering" to encompass the wider scope of the field, and actively engages in regulatory affairs and work on documents relevant to the MBES community.

As anticipated, MBES are major enablers of modern medicine, which has made its main advancements in the past decades mainly because of the main output of BME. The MBES community has direct impact on quality of life, safety, and cost-effectiveness of healthcare services, with our effort in research, innovation, assessment, procurement, and management of BME main products, which are biotechnologies and medical technologies (MedTech) such as medical devices (MDs), personal protective equipment (PPE), and digital health.

The EAMBES is also affiliated to the International Federation of Medical and Biological Engineering (IFMBE)[11], which is member of the International Union for Physical and Engineering Sciences in Medicine (IUPESM)[12]. The EAMBES is an active member of Health First Europe (HFE)[13] which is a non-profit, non-commercial alliance of patients, healthcare workers, academics, healthcare experts, and the medical technology industry who are joining forces to transform healthcare through innovative solutions.

The EAMBES is a membership organization with over 100 members, including national and regional scientific societies, academic and research institutions, and individual members. This means that this alliance emphasizes the importance of a unified voice for MBES in Europe due to the field's diversity and fragmentation, and works across research, education, and healthcare sectors, recognizing the key role of MBES professionals in the European medical technology industry.

The benefits of being a member of the EAMBES is the access to a network of MBES professionals from all over Europe, the augmented opportunities to participate in research projects, the information on the latest developments in MBES and much more.

Therefore, the EAMBES tackles the crucial task of propelling the field of medical and biological engineering and science forward through a holistic approach. Firstly, it champions the significance of MBES research and development by lobbying policymakers and stakeholders, while also injecting financial support into research projects and nurturing future talent through fellowships. Secondly, the EAMBES recognizes the importance of fostering knowledge within the field, and actively collaborates with universities and educational institutions to refine and expand MBES education and training programs. Additionally, it equips educators and students with valuable resources. Finally, the EAMBES understands the power of collaboration, and cultivates a platform for MBES professionals from around the globe to exchange information and join forces on research projects. By facilitating conferences and workshops, it fosters a vibrant and interconnected MBES community, paving the way for shared discoveries and advancements in the field.

The EAMBES and European Parliament and Commission: A Fruitful Collaboration for the Future of Healthcare

During the last decade, the EAMBES has forged a stronger connection with the European Parliament, leveraging its expertise to shape healthcare policy and advance innovation. This collaboration, spearheaded by the Policy Affairs Working Group (PAWG), promises a brighter future for European healthcare, with a focus on tackling pressing challenges and promoting cutting-edge solutions. The EAMBES is committed to supporting the European Parliament and Commission in shaping current and future health policies. This includes improving regulations on MDs, PPE, and artificial intelligence (AI), along with addressing emerging challenges like cybersecurity and emergency preparedness. Fostering research investments in fields like aging, cardiovascular diseases, and the European Union (EU) dataspace are also key priorities, aiming to leverage AI for healthcare advancements. By advocating for the field's recognition, and promoting strategic autonomy, the EAMBES is committed to ensuring that Europe remains at the forefront of healthcare advancements, benefiting citizens across the continent and beyond.

The journey began on 2016 with the establishment of the European Parliament Interest Group (EPIG) on Biomedical Engineering (BME). The first EPIG[14] meeting was successfully held in Brussels with the support of IFMBE. Interesting presentations were followed by lively discussions and the action points were excellent first steps towards full recognition of BME in European policy and funding instruments. This marked a significant step towards recognizing the field's importance in policymaking.

In 2018, the PAWG organized the second meeting of the EPIG BME, in collaboration with the IFMBE and the World Health Organization (WHO), and with representatives of the African community of BMEs. The meeting's main focus with the European Parliament was to achieve a crucial step forward in advocating for the advancement of BME. Firstly, the meeting emphasized the need to formally recognize BME as an independent profession. This includes listing it within the European Skills, Competences, Qualifications and Occupations (ESCO) database, a vital tool for employment and skills identification. Such recognition would not only ensure proper identification of BME professionals but also contribute to their career development and mobility within the European Union. Secondly, the need for dedicated research space within the European Research Council (ERC), Future and Emerging Technologies (FET), and Marie Skłodowska-Curie Actions (MSCA) was highlighted. Creating a dedicated BME panel within these schemes would significantly boost research efforts in this rapidly evolving field, leading to groundbreaking innovations in healthcare and medical technology. Finally, the meeting emphasized the importance of establishing a stable collaborative relationship with the African Union. This collaboration could help harmonize regulations on MDs and medical locations across Africa, ensuring the safety and well-being of African citizens while also offering valuable opportunities for European MD manufacturers, particularly small- and medium-sized enterprises (SMEs). By addressing these crucial points, the meeting took a significant step towards fostering a more robust and impactful BME ecosystem in Europe and beyond. Recognizing BME as an independent profession, supporting dedicated research initiatives, and fostering collaboration with Africa presents exciting opportunities for advancements in healthcare and a brighter future for all involved[15].

The third meeting with European Parliament Members was held online in 2022, entitled "*Practices, Impact and Promises of Biomedical Engineering for advancing the EU's Health Priorities*" and aimed at restoring the EPIG BME[16]. The webinar focused on highlighting the role of BME in the EU healthcare landscape and the impact that certain policies and initiatives can have on tackling pressing issues. Panelists elaborated on how biomedical engineering, MDs, and health data can come together and work towards solutions that directly benefit patients. The EAMBES has pledged its continued support to the European Parliament and Commission in shaping healthcare policies for the future. Recognizing the need for robust *regulations*, the EAMBES advocated for improvements in areas MDs, PPE, and AI in healthcare. They emphasized the importance of addressing emerging challenges such as cybersecurity, emergency preparedness, and domestic production of essential medical supplies. Additionally, they called for increased autonomy in digital infrastructure and service providers. To ensure effective resource allocation, the EAMBES highlighted the need to prioritize research and innovation investments in areas of critical need, such as ageing and cardiovascular diseases. Furthermore, they advocated for the development of an EU-wide dataspace to facilitate the growth of responsible AI applications in healthcare. Through these combined efforts during the EPIG BME events, the EAMBES aims to ensure a more resilient, innovative, and secure healthcare landscape for all Europeans.

Building on the momentum, the EAMBES co-organized in 2023, a dedicated hybrid event at the European Parliament entitled: "*Pandemic Management and Preparedness*"[17]. This insightful discussion explored the role of telemedicine, innovative technologies, and global cooperation in ensuring a safer future[18]. The event aimed to pave the way for experts and policymakers to strengthen their dialogue, ensure collaboration between stakeholders, and ensure that the EU will exit the pandemic stronger, more resilient, and more autonomous than before. In the first part, the event aimed to identify the "lessons learned" from the pandemic, in regard to telemedicine and other health technologies used during this period, in order to highlight various best practices followed. In the second panel, the event yielded the floor to medical technology experts in order to explore efficient measures to prepare for the next health crisis in Europe. Throughout these interactions, the EAMBES has consistently positioned itself as the preeminent voice for BME in Europe, consistently pushing for advancements that benefit both the field and society as a whole. They champion independent professional recognition, solidifying BME practitioners as key figures in healthcare. By advocating for dedicated research funding, the EAMBES seeks to accelerate innovation and tackle pressing healthcare challenges. Additionally, they call for harmonized regulations across Europe and Africa, smoothening collaboration and the flow of medical technologies. Recognizing the strategic importance of MD production, the EAMBES promotes European autonomy in this field, ensuring healthcare resilience and timely access to critical technologies. Finally, they envision unlocking the power of AI in healthcare by enabling the EU dataspace, paving the way for personalized medicine and improved patient outcomes. Through these efforts, the EAMBES champions a vision of a strong, innovative, and patient-centered BME landscape in Europe and beyond.

The collaboration between the EAMBES and the European Parliament is not merely a historical record; it represents a blueprint for the future. By harnessing the power of BME, we can build a healthier, more resilient Europe, equipped to tackle current and future health challenges. This journey requires continued commitment from policymakers, researchers, and healthcare professionals alike. By working together, we can ensure that innovation remains at the forefront of healthcare, delivering a brighter future for all.

The Role of the EAMBES in Involving African BME in the European Context

The EAMBES plays a crucial role in facilitating the involvement of African BMEs in the European context. The EAMBES advocates for African voices, fosters knowledge sharing, and builds a network that empowers collaboration and individual recognition. One key aspect is policy engagement. The EAMBES collaborates with the European Parliament, ensuring African BME perspectives are heard in shaping regulations and research frameworks like Horizon Europe. This not only benefits the African BME community but also enriches the entire field with diverse viewpoints.

Furthermore, the EAMBES fosters international collaboration. By working with organizations like the IFMBE and WHO, the EAMBES organizes events that bring together BME communities from across continents. This dialogue paves the way for partnerships and knowledge exchange. Building capacity is another crucial area. The EAMBES supports educational and training programs in Africa, working with institutions to develop relevant curricula and exchange programs. This increases African BME participation in European educational opportunities, fostering a new generation of skilled professionals. Knowledge sharing is central to the EAMBES' mission. They organize conferences, workshops, and webinars featuring African BME expertise, bridging the gap between African and European researchers. This platform empowers African BME professionals to showcase their work and contribute to the advancement of the field.

Networking and collaboration are also cornerstones of the EAMBES' work. They encourage the formation of BME organizations in Africa, strengthening regional networks and paving the way for stronger connections with the European BME community. Additionally, the EAMBES recognizes individual and group achievements, creating role models and inspiring further engagement.

While the EAMBES plays a significant role, there are still challenges to overcome, such as limited funding, language barriers, and differing research priorities. Access to funding and limited infrastructure in Africa can hinder participation. The EAMBES advocates for funding opportunities and supports joint research projects to address these issues. By bridging these gaps, the EAMBES empowers African BME professionals, fostering a more inclusive and collaborative global BME landscape. This, in turn, benefits not only Africa and Europe but also the entire field of biomedical engineering, leading to advancements that improve lives worldwide.

Overall, the EAMBES plays a vital role in bridging the gap between African and European BME communities by advocating for their voices, facilitating knowledge sharing, fostering collaboration, and addressing existing challenges. This ultimately benefits both the European and African BME communities, leading to advancements in the field and improved healthcare outcomes globally.

The Role of the EAMBES During COVID-19

During the years of the COVID-19 pandemic, the EAMBES emerged as a crucial player in Europe's response. Leveraging its unique position as a bridge between policymakers and the vast biomedical engineering community, the EAMBES facilitated essential dialogue and collaboration, propelling innovation towards solutions for the crisis and the urgent needs of public health. Leveraging its network of experts and member societies, the EAMBES fostered collaboration and knowledge sharing across Europe, accelerating the development and deployment of critical technologies.

Secondly, the EAMBES acted as an advocate for the field, engaging with policymakers at the European and national level to ensure adequate funding and resources were directed towards these critical endeavors[19]. This advocacy helped accelerate research and development timelines, ultimately contributing to the global fight against the pandemic. By bridging the gap between science and policy, the

EAMBES played a significant role in harnessing the power of BME to mitigate the impact of COVID-19 and pave the way for future preparedness.

While the pandemic posed unprecedented challenges, the EAMBES's role in facilitating collaboration, advocating for the field, and connecting research to real-world needs proved invaluable. Its efforts not only contributed to the fight against COVID-19, but also laid the groundwork for a more resilient and responsive healthcare system in the face of future global health threats. While the EAMBES' activities remain ongoing, its efforts during COVID-19 stand as a testament to the power of collaboration and knowledge sharing in the face of global challenges. The lessons learned and connections forged during this period continue to shape the alliance's work, ensuring its continued contribution to a healthier and more resilient Europe.

Once again, the EAMBES' contribution to the global effort against COVID-19 highlights the importance of strong alliances within scientific communities, particularly in times of crisis. As the world continues to navigate evolving health challenges, the EAMBES' role in fostering collaboration and advocacy within the field of BME remains crucial.

The EAMBES Suggestions for Facing Healthcare Emergencies in Low-Resource Settings

The EAMBES recognizes the unique challenges faced by low-resource settings (LRSs) during healthcare emergencies and advocates for a multipronged approach focused on both prevention and preparedness, and technology-driven solutions adapted to local contexts. Prevention and preparedness involve strengthening local healthcare systems through education and training of healthcare professionals, including biomedical engineers who can maintain and repair medical equipment. Early warning systems and community engagement are crucial to mobilize resources and respond swiftly to outbreaks. The EAMBES also emphasizes the importance of affordable and accessible diagnostics and treatments tailored to specific diseases prevalent in low-resource regions. Technology plays a vital role in the EAMBES' suggestions and specifically promotes the development of low-cost, point-of-care diagnostic tools, telemedicine solutions for remote consultations, and mobile health platforms for data collection and disease surveillance. Additionally, the EAMBES encourages open-source MD designs and local manufacturing capabilities to ensure sustainability and affordability.

In essence, the EAMBES' suggestions prioritize resourcefulness, innovation, and empowerment to equip LRSs with the tools and knowledge they need to navigate healthcare emergencies effectively. By combining collaboration and technological innovation, the EAMBES believes that we can build more resilient healthcare systems in LRSs, ultimately improving health outcomes and saving lives during emergencies.

Notes

1 L. Pecchia, *Health technology assessment* in E. Iadanza, *Clinical Engineering Handbook*, Second Edition Elsevier, 2019, pp. 787–788.
2 R. Castaldo, C. Federici, L. Pecchia, *Early-stage healthcare technology assessment*, in E. Iadanza (ed by), *Clinical engineering handbook.* Second edition, cit., pp. 799–806, ivi, p. 799.
3 EUR-LEX, 2015/C 291/07: EUR-LEX, 2015/C 291/07, "Opinion of the European Economic and Social Committee on Promoting the European single market combining biomedical engineering with the medical and care services industry", Rapporteur: Edgardo Maria Iozia, Co-rapporteur: Dirk Jarré cit. in L. Pecchia, et al., *Health technology assessment teaching for BME*, in *Clinical Engineering Handbook*, cit., p. 832–835, ivi, p. 832.
4 L. Pecchia et. al, *Health technology assessment teaching for BME*, in *Clinical Engineering Handbook*, cit., p. 832.
5 L. Pecchia et al., *Harmonising medical device and medical location policies among Africa and Europe*, in *Proceedings of the fourth WHO global forum on medical devices*, Visakhapatnam, India, AMTZ-Kalam Convention Centre (2018).
6 L. *Pecchia, N. Pallikarakis, R. Magjarevic, E. Iadanza E., Health Technology Assessment and Biomedical Engineering: Global trends, gaps and opportunities, «Medical Engineering and Physics»,* 7 (2019), pp. 19–26.
7 European Alliance for Medical and Biological Engineering and Science: State of the alliance 2021: https://eambes.org/wp-content/uploads/2022/02/EAMBES-2021_StateOfTheAlliance.pdf, p. 4.
8 WHO, *Human resources for medical devices. The role of biomedical engineers*, WHO Medical Devices Technical Series, Ginevra, 2017.
9 A. Pacela, *Bioengineering Education Directory*, Brea, Calf, 1990.
10 https://eambes.org/
11 https://ifmbe.org/
12 https://iupesm.org/
13 https://healthfirsteurope.eu/
14 https://eambes.org/the-first-meeting/
15 https://eambes.org/the-second-meeeting/
16 https://eambes.org/the-third-meeting/
17 https://eambes.org/the-fourth-epig-meeting/
18 A. Maccaro, C. Audia, K. Stokes, H. Masud, S. Sekalala, L. Pecchia, D. Piaggio, *Pandemic preparedness. A scoping review and guidelines for the future*, in «MDPI Healthcare», Vol. 11. No. 18. 2023, pp. 1–20: www.mdpi.com/2227-9032/11/18/2572
19 A. Maccaro, D. Piaggio, C. Dodaro, L. Pecchia, *Biomedical engineering and ethics: reflections on medical devices and PPE at the time of Covid-19*, in «BMC Medical Ethics», (2021) 22:130 https://doi.org/10.1186/s12910-021-00697-1

Chapter 5

Conclusions

The COVID-19 pandemic ripped through our world, exposing gaping inequalities in access to healthcare, especially in resource-scarce regions. This book, a testament to the vital role of collaboration and knowledge sharing in addressing global health challenges, has illuminated the crucial intersection of ethics, biomedical engineering (BME), and their application in diverse settings. As we navigate the aftermath of the COVID-19 pandemic and brace for future global health emergencies, this concluding chapter aims to synthesize the key takeaways and offer a roadmap for navigating future challenges.

The pandemic starkly highlighted the ethical challenges posed by resource scarcity. Decisions made by healthcare workers in the face of limited supplies echoed throughout the book's chapters, prompting reflection on the inadequacy of established regulatory frameworks in times of crisis. As the articles on medical devices (MDs) and personal protective equipment (PPE) (Maccaro et al., 2021; Pecchia et al., 2020) reveal, healthcare workers faced agonizing choices amidst limited supplies. The need for rapid innovation clashed with established regulatory frameworks, exposing their inadequacy in times of crisis. This underscores the crucial role of flexible and context-sensitive ethical frameworks that can guide decision-making in resource-constrained environments.

Moving forward, fostering closer partnerships between researchers, engineers, policymakers, and communities, particularly in low-resource settings (LRSs), is imperative. Continued investment in research and development tailored to the specific needs of these communities is crucial for building resilient healthcare systems that can withstand future pandemics and other global health emergencies. The articles on LRSs (Maccaro et al., 2022a, 2022b; Piaggio et al., 2021; Di Pietro et al., 2020) remind us that healthcare interventions cannot be divorced from the social and cultural realities of the communities they aim to serve. Community engagement, cultural sensitivity, and respect for local knowledge are not just ethical imperatives but also practical prerequisites for successful interventions. The Benin case studies (Maccaro et al., 2022a, 2022b) vividly illustrate this point, showcasing how social

DOI: 10.1201/9781003436348-14

dynamics and local beliefs can influence the acceptability and effectiveness of healthcare programs.

The universality of medical device regulations (MDRs) (Maccaro et al., 2022c) is challenged by the vastly different realities of high- and low-income countries (HICs and LICs). A one-size-fits-all approach can stifle innovation and hinder access to essential technologies in resource-limited settings. The call for context-specific regulations and ethical guidelines resonates throughout the book, urging us to move beyond a Eurocentric approach and embrace diversity in healthcare solutions.

The rapid advancement of technologies like artificial intelligence (AI) brings immense benefits and exciting possibilities, but also raises complex ethical concerns. Navigating these considerations requires ongoing dialogue and collaboration among scientists, policymakers, ethicists, and the public. Building robust ethical frameworks, promoting responsible development and use of new technologies, and ensuring public trust are essential for harnessing their full potential while mitigating potential harms.

Despite its immense potential, the evolving role of BME faces challenges. Ethical considerations surrounding data privacy, access to technology, and potential inequities in its application require careful navigation. Collaboration across disciplines, including public health, social sciences, and policymakers, is crucial to ensure responsible and equitable implementation of these technologies. The future of BME in addressing global health challenges is bright. By leveraging its innovative power and focusing on collaborative, ethical, and sustainable solutions, this field can play a pivotal role in creating a healthier and more equitable world for all.

Emerging from the pandemic's shadow, we must embrace adaptability in healthcare systems to respond to evolving needs, foster open dialogue for ethical technology implementation, and prioritize inclusivity by designing interventions with context and equity in mind. By embracing these principles and actively supporting local researchers and practitioners in low- and middle-income countries (LMICs), we can build local capacity, foster research addressing their specific needs, and advocate for equitable access to healthcare as a fundamental human right, contributing to a just and sustainable future for all.

Emerging from the pandemic's shadow, the lessons learned reverberate. We must embrace adaptability, ensuring healthcare systems bend with the wind of evolving needs, particularly in crisis. Open dialogue and transparency are vital, building trust and paving the way for the ethical implementation of new technologies. Above all, inclusivity reigns supreme. Healthcare interventions must be crafted with context in mind, respecting local realities and diligently addressing inequities. These are the pillars upon which we build responsive ethical frameworks, contributing to a future where justice and sustainability flourish for all.

Yet, for low- and middle-income countries (LMICs), ethically sound solutions demand more than technological leaps and well-meaning intentions. Active support and empowerment of local researchers and practitioners are paramount. Building robust research, development, and regulatory infrastructure within LMICs is crucial for ensuring equitable access to safe and effective technologies. Investing in local researchers, fostering their participation in global health initiatives, and advocating relentlessly for equitable access are not mere actions, but cornerstones in the fight for healthcare as a fundamental human right for all.

The COVID-19 pandemic may be a chapter closed, but the story of global health emergencies continues. "Ethics and Biomedical Engineering Face Global Health Emergencies" serves not just as a record of the past, but as a springboard for the future. It reminds us that ethics, biomedical engineering, and global health are not separate strands; they are interwoven threads in the tapestry of a healthier and more just world. Let us not be defined by the scars of the past, but by the bridges we build together. Let us remember, the pandemic may have dimmed the light, but the lessons learned illuminate the path towards a future whereby embracing the ethical considerations, technological innovations, and lessons learned from this unprecedented crisis, we can build a more resilient and unbiased healthcare system that is better prepared to face the challenges of the future. Let us remember that the ethical imperative to protect and care for all, particularly the most vulnerable, must guide our response to global health emergencies, both now and in the years to come.

Index

B

C

D

E

F

G

H

I

J

W

Y

Z

For Product Safety Concerns and Information please contact our EU
representative GPSR@taylorandfrancis.com
Taylor & Francis Verlag GmbH, Kaufingerstraße 24, 80331 München, Germany

www.ingramcontent.com/pod-product-compliance
Ingram Content Group UK Ltd.
Pitfield, Milton Keynes, MK11 3LW, UK
UKHW022125190926
475702UK00013B/1644